Key Stage 3 Science

Spectrum 8

Andy Cooke

Jean Martin

CAMBRIDGE
UNIVERSITY PRESS

Series editors	Andy Cooke
	Jean Martin
Physics consultant	Sam Ellis
Chemistry consultant	Doug Wilford
Authors	Darren Beardsley
	Jenifer Burden
	Chris Christofi
	Zoe Crompton
	Sam Ellis
	David Fagg
	Jean Martin
	Janet McKechnie
	Mick Mulligan
	Nicky Thomas

PUBLISHED BY THE PRESS SYNDICATE OF THE UNIVERSITY OF CAMBRIDGE
The Pitt Building, Trumpington Street, Cambridge, United Kingdom

CAMBRIDGE UNIVERSITY PRESS
The Edinburgh Building, Cambridge CB2 2RU, UK
40 West 20th Street, New York, NY 10011-4211, USA
477 Williamstown Road, Port Melbourne, VIC 3207, Australia
Ruiz de Alarcón 13, 28014 Madrid, Spain
Dock House, The Waterfront, Cape Town 8001, South Africa

http://www.cambridge.org

© Cambridge University Press 2003

First published 2003

Printed in the United Kingdom by Cambridge University Press

Typeface Delima MT   *System* QuarkXPress®

A catalogue record for this book is available from the British Library

ISBN 0 521 75007 5   paperback

Cover design by Blue Pig Design Co
Page make-up and illustration by hardlines Ltd, Charlbury, Oxford

Contents

About Spectrum

This *Spectrum* Class Book covers what you will learn about science and scientists in Year 8. It is split into twelve **Units**. Each Unit starts with page a like this:

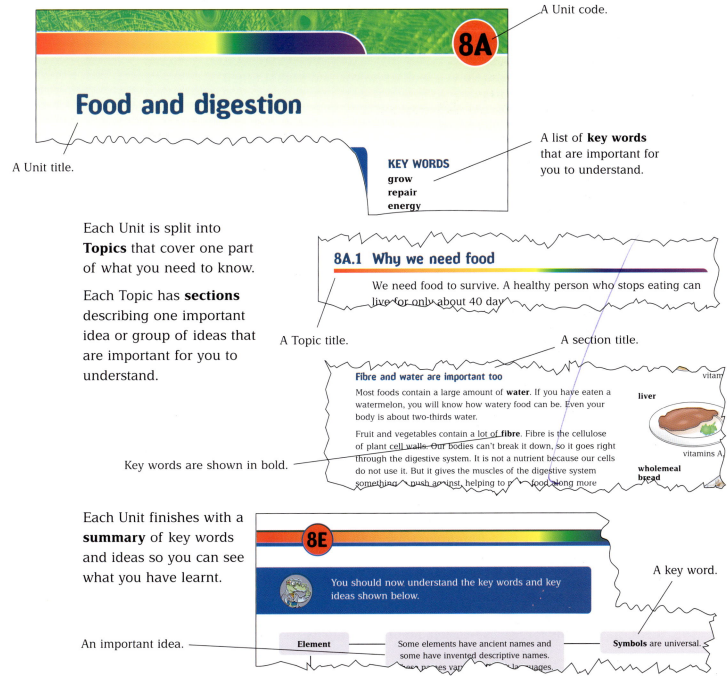

A Unit code.

Food and digestion

8A

A Unit title.

KEY WORDS
grow
repair
energy

A list of **key words** that are important for you to understand.

Each Unit is split into **Topics** that cover one part of what you need to know.

Each Topic has **sections** describing one important idea or group of ideas that are important for you to understand.

8A.1 Why we need food

We need food to survive. A healthy person who stops eating can live for only about 40 day

A Topic title.

A section title.

Fibre and water are important too

Most foods contain a large amount of **water**. If you have eaten a watermelon, you will know how watery food can be. Even your body is about two-thirds water.

Fruit and vegetables contain a lot of **fibre**. Fibre is the cellulose of plant cell walls. Our bodies can't break it down, so it goes right through the digestive system. It is not a nutrient because our cells do not use it. But it gives the muscles of the digestive system something to push against, helping to move food along more

Key words are shown in bold.

vitam

liver

vitamins A

wholemeal bread

Each Unit finishes with a **summary** of key words and ideas so you can see what you have learnt.

8E

You should now understand the key words and key ideas shown below.

A key word.

An important idea.

Element

Some elements have ancient names and some have invented descriptive names.

Symbols are universal.

Icons

Introducing each Unit and the summary of key words and ideas at the end of each Unit.

Asking questions about what you have just learnt and telling you where to look in the Class Book to help with activities.

Asking questions that help you think about what you have just learnt.

Asking questions sthat might need some research to answer.

At the end of the book

At the end of the book you will find:

- pages 147 to 152 to help you with **scientific investigations**.

- a **glossary/index** to help you look up words and find out their meanings.

Other components of Spectrum.

Your teacher has other components of *Spectrum* that they can use to help you learn. They have:

- A **Teacher File** or **Teacher CD-ROM** full of information for them and lots of activities of different kinds for you. The activities are split into three levels: **support**, **main** and **extension**. Some of the activities are **suitable for homework**;

- an **assessment CD-ROM** with an **analysis tool**. The CD-ROM has **multiple choice tests** to find out what you know before you start a Unit and for you to do during or after a Unit. It also has some end of year **SAT-style tests**;

- a set of **Technician Notes** with information about **practical activities**;
  (and free on the web available at www.cambridge.org/spectrum)

- general guidance documents on aspects of the Science Framework;

- **investigation checklists**, **investigation sheets** – writing frames to help with structuring investigations, and **level descriptors** covering **Planning**, **Observation**, **Analysis**, **Evaluation** and **Communication**;

- **mapping grids** for the **Five Key Ideas**, **Numeracy**, **Literacy**, **ICT**, **Citizenship** and **Sc1**;

- **flash cards** for use as revision aid or for card chases using the Year 8 key words;

- **Five Key Ideas cards** for use as a revision aid and to build giant concept maps.

Food and digestion

In this unit we shall be learning about different food groups and what we need for a balanced, healthy diet. We shall also be learning about how our bodies break down food and use it for growth, repair of damage, and as an energy source.

KEY WORDS
grow
repair
energy
nutrients
proteins
carbohydrates
fats
vitamins
minerals
water
fibre
absorb
digestion
faeces
enzymes

8A.1 Why we need food

We need food to survive. A healthy person who stops eating can live for only about 40 days.

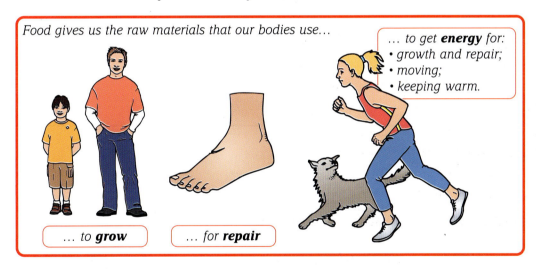

Food gives us the raw materials that our bodies use...

... to get **energy** for:
• *growth and repair;*
• *moving;*
• *keeping warm.*

... to **grow**   ... for **repair**

1   Look at the pictures. Write down:

    a <u>two</u> reasons you need to make new cells;

    b <u>three</u> uses for energy in your body.

We call the food substances that our cells use **nutrients**. They are:

● **proteins**, for making new cells;

● **carbohydrates** and **fats** for energy;

● small amounts of **vitamins** and **minerals**.

2   Find out a use of <u>one</u> vitamin and <u>one</u> mineral in your body.

We are what we eat

Proteins are the main raw materials for making new cells. So proteins are particularly important at times when we are growing quickly.

Taribo has a healthy diet.

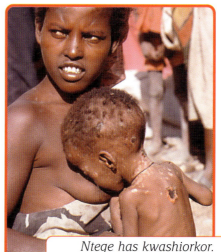

Ntege has kwashiorkor. Kwashiorkor is a disease caused by a lack of proteins in his diet.

3   Write down <u>two</u> differences between the children in the photographs.

4   Write down <u>two</u> foods that will improve Ntege's health if he can get them.

5   Some children don't eat enough protein foods. Will this have any long-term effects on them? Explain your answer.

6   On average, a pregnant woman needs 76g of protein per day. A woman who is not pregnant needs less.

   a Explain why a pregnant woman needs extra protein.

   b Find out how much protein a woman normally needs.

When we cut ourselves some cells are damaged, some die and others are lost when we bleed. Our bodies have to make more cells to repair the wound and to replace the lost and damaged cells. Cells in our bodies are continually dying and being replaced by new ones. For example, a red blood cell lasts for only about four months. New red blood cells are made all the time to replace those that are worn out.

Look at the photographs. The wound is gradually healing up. Eventually, there will be no sign of it.

George's finger after an accident

George's finger is healing up

7 Explain what is happening to make the wound on George's finger heal up.

8 Wounds take longer to heal if the person doesn't eat enough protein foods. Why is this?

We eat for energy too

We need energy for:

- growth and repair of cells;

- moving;

- keeping warm.

We need raw materials <u>and</u> energy from our food to make and repair cells. So energy as well as proteins are important when we are growing. We release this energy from the carbohydrates and fats in our food.

Carbohydrate foods

starchy

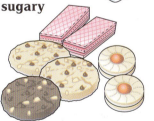

sugary

9 Write down <u>four</u> energy foods that you eat.

10 Lack of energy foods affects a child's growth. Why is this?

11 Explain the effects of a lack of energy foods on adults who are no longer growing.

Fatty foods

Muscles contract to make us move. To contract, muscles need energy. So, the more we move around, the more energy we need. Even when we are not moving around, our hearts are beating and other muscles are helping us to breathe and to keep food moving through our digestive system.

The table shows how much energy we need for different activities. We measure energy in kilojoules (kJ).

Activity	kJ/hour
Sitting	63
Standing	84
Walking	750
Swimming	1800
Walking upstairs	4184
Sprinting	5183

12 When we are sitting still, what do our bodies use energy for?

13 Between which <u>two</u> activities in the table does a person getting dressed fit? Explain your answer.

14 Adam is a distance runner. On the day before a race, he eats lots of carbohydrate foods.
Why do you think he needs to do this?

15 Why do we need more energy when standing up than sitting down?

Mini but mighty

We also need small amounts of vitamins and minerals. Although the amounts that we need are very small, they are very important for our health.

In the 1740s two-thirds of sailors died from a disease called scurvy. When they were away from land for a long time, they didn't eat any fresh fruit or vegetables. So they didn't have any vitamin C in their diet, and lack of vitamin C causes scurvy.

Minerals such as calcium and iron are also important.

Calcium is a raw material for making bones and teeth, and iron is needed to make red blood cells.

Scurvy causes bleeding gums.

16 Look at the photograph. Describe some effects of scurvy.

17 Write down <u>one</u> food that contains:

    a vitamin C;

    b vitamins A and D and the mineral iron;

    c calcium.

The label shows some of the nutritional information from a packet of Sugary Puffs.

18 What is the main nutrient in Sugary Puffs?

19 Which mineral is found in Sugary Puffs?

20 Write down <u>one</u> vitamin in Sugary Puffs.

21 Find out the effects of lack of each of the two B vitamins.

NUTRITIONAL INFORMATION	
TYPICAL VALUE per 100 g	
Energy	1620 kJ
Protein	6.5 g
Carbohydrates	86.5 g
(of which sugars)	49.0 g
Fat	1.0 g
Fibre	3.0 g
VITAMINS	
Thiamin (B1)	1.0 mg
Riboflavin (B2)	1.0 mg
MINERALS	
Iron	8.0 mg

Foods containing vitamins and minerals

fish

vitamins A, D

milk

vitamins A, D; calcium

vegetables

vitamins A, B, C

egg

vitamins B, D

liver

vitamins A, D; iron

wholemeal bread

B vitamins, iron, calcium

Fibre and water are important too

Most foods contain a large amount of **water**. If you have eaten a watermelon, you will know how watery food can be. Even your body is about two-thirds water.

Fruit and vegetables contain a lot of **fibre**. Fibre is the cellulose of plant cell walls. Our bodies can't break it down, so it goes right through the digestive system. It is not a nutrient because our cells do not use it. But it gives the muscles of the digestive system something to push against, helping to move food along more easily. Without it, you'd be very constipated! In fact, fibre and water make up a large part of the bulk of your food.

22 Explain why you need water and fibre in your diet.

8A.2 A healthy diet

A balanced, healthy diet contains the correct amount of each food group. We can get a balanced diet in all sorts of ways. Many people in richer countries like the USA and the UK get a lot of their protein from meat. Most people in poorer countries like India and China rely on cereals and beans for their proteins.

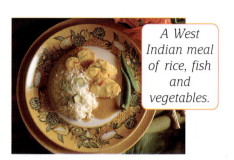

A West Indian meal of rice, fish and vegetables.

1 In the West Indian meal, the rice is the main energy source. What provides most of the protein?

2 In the Chinese dish, the prawns provide most of the protein. What is the main energy source?

3 Which part of the European meal contains most of the proteins and fat?

A Chinese meal of noodles, prawns and vegetables.

Every meal that you eat does not have to be balanced. It is what you eat over several meals that matters. You can find out the amount of each group in a packaged food by looking at the label.

A European meal of meat, potato and vegetables.

4 Which of the foods on the graph contains the most:

 a water?   b protein?

5 What is the main nutrient in potatoes?

6 Which nutrients in the graph are missing from cod?

An analysis of the main nutrient content, water and fibre in four foods

A healthy, balanced diet is different for different people. The amount of each nutrient we need depends on:

- our age;
- whether we are male or female;
- our body size;
- the activities and jobs that we do.

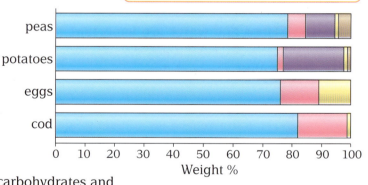

A person doing building work needs more carbohydrates and fats for energy than a person working at a desk all day.

Key
- water
- fibre
- fat
- carbohydrate
- protein

7 What do the following people need to eat?

 a Mmapula, a 13-year-old girl living in South Africa.

 b Steve Carr, who plays professional football for Spurs.

8 Janet is breastfeeding her baby. Find out what she should eat and drink.

8A.3 Getting nutrients out of your food

The nutrients in your food have to pass into your blood. We say that you **absorb** them. The particles of vitamins, minerals and some sugars such as glucose are small enough to be absorbed. The large, insoluble molecules of fats, proteins and some carbohydrates are not. So you have to break them down into smaller molecules. We call this process **digestion**. It happens in your digestive system. After digestion, the small molecules pass into your blood and are transported to your cells.

1 Write down <u>three</u> substances that you can absorb without digesting them.

Modelling what happens in your digestive system

In science, we sometimes use models to help us understand how things work. We can use this model of the gut to find out which substances can pass into the blood and which can't.

2 What part of the diagram represents the blood?

3 What does the visking tubing represent?

4 Later, there is glucose in the water around the visking tubing, but no starch. Explain why.

starch and glucose solutions

water

visking tubing

boiling tube

Look at the diagram of what happens in your gut.

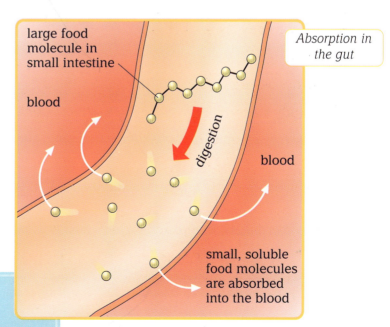

large food molecule in small intestine

blood

digestion

blood

small, soluble food molecules are absorbed into the blood

Absorption in the gut

5 Which kind of molecules:

a can pass into your blood?

b cannot pass into your blood?

Explain your answers.

Your digestive system

The food's journey through your digestive system starts in your mouth and ends when it passes out through your anus as **faeces**. The journey is 8 to 9 metres long and usually takes between 24 and 48 hours to complete.

If food goes through too quickly, it is not broken down into the nutrients that you need. If the surface area of your digestive system is not large enough, you will not be able to absorb all the nutrients.

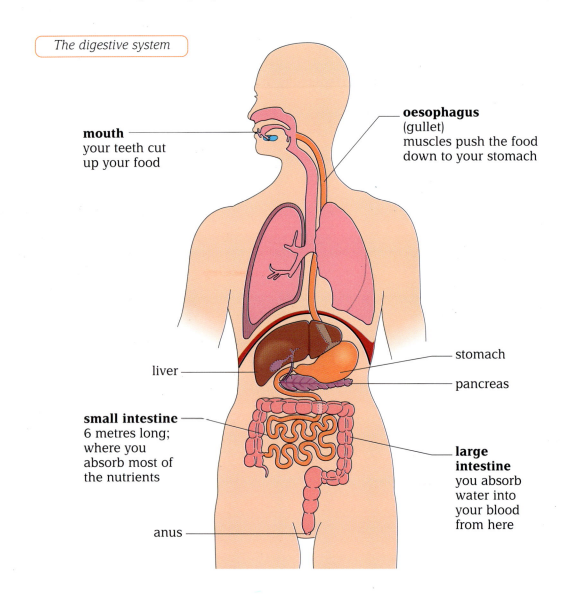

The digestive system

mouth
your teeth cut up your food

oesophagus
(gullet)
muscles push the food down to your stomach

liver

stomach

pancreas

small intestine
6 metres long; where you absorb most of the nutrients

large intestine
you absorb water into your blood from here

anus

6 List, in order, the parts of your digestive system that your food travels through.

8A.4 How your digestive system works

Over 3000 years ago, the Ancient Greeks described many of the organs of the digestive system. They found out about the different organs by dissecting dead bodies, sometimes in public places.

In the 1760s, an Italian priest called Lazzaro Spallanzani did experiments on his own body to find out about the digestion of food. He swallowed wooden blocks with holes containing meat and collected them when they passed out of his anus. He discovered that the food in the wooden blocks had disappeared. Spallanzani also made himself vomit and showed that the liquid vomit dissolved away meat. Sometimes he swallowed food on a piece of thread and pulled the food out before it was fully digested.

Lazzaro Spallanzani

1 a What happened to the meat in the wooden blocks that Spallanzani swallowed?

   b Where did this happen?

We now know that Spallanzani's meat disappeared because **enzymes** had broken it down. We know that cells in some parts of our digestive systems release these chemicals and that they break down large molecules of food into smaller ones.

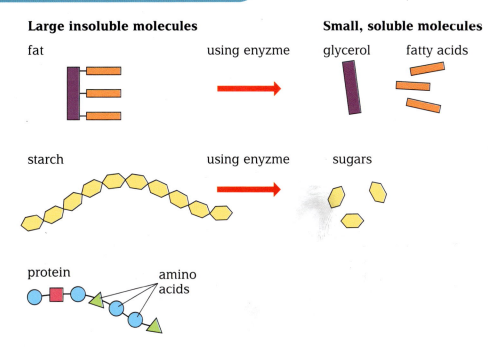

Large insoluble molecules

fat

starch

protein

Small, soluble molecules

using enyzme

glycerol   fatty acids

using enyzme

sugars

amino acids

2 Write down the names of the molecules produced by the breakdown of:

   a fat;

   b starch.

3 Proteins break down into amino acids. Draw a diagram to show what the protein in the diagram looks like when it is broken down.

Now look at the gut model again.

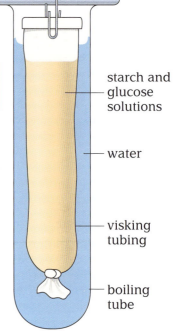

starch and glucose solutions

water

visking tubing

boiling tube

4 What kind of substance can you put in the visking tubing to break down the starch into sugar?

5 What happens to the sugar that is made?

6 Your saliva breaks down starch. What does this tell you about your saliva?

7 Think carefully of ways in which the model is different from a real small intestine. Write down your answers.

Scientists continue to research what happens to food in the digestive system. They have found out that different enzymes break down different foods, and that different enzymes work best in different conditions. For example, some work best in acidic conditions, others in alkaline.

Often the research is done to find out more about illnesses. One way is to give a patient a harmless liquid containing a barium compound to drink. The barium shows up on an X-ray photograph as it moves through the digestive system. Doctors can even look inside the stomach using an endoscope.

A tiny camera on the tip of an endoscope can photograph inside the digestive system.

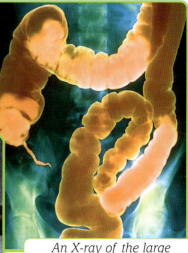

An X-ray of the large intestine after a barium meal

8 Write down <u>two</u> things that scientists have found out about enzymes.

9 Find out:

a the connection between an endoscope, fibre optics and a stomach ulcer;

b <u>one</u> reason why doctors take X-rays of a patient's digestive system.

8A.5 After digestion

Every cell of the body needs nutrients. Cells need them for growth, repair and as an energy source. So the bloodstream carries the nutrients absorbed in the small intestine to all parts of the body. They are carried in solution in the blood plasma – the liquid part of blood.

1 Write a list of nutrients that can pass into your blood.

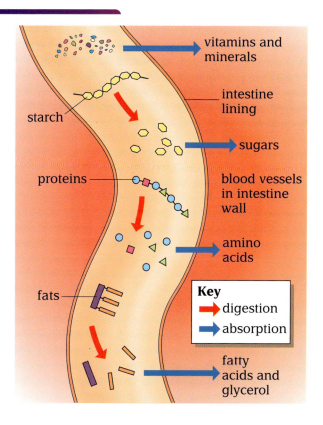

Bloodstream   **Liver**

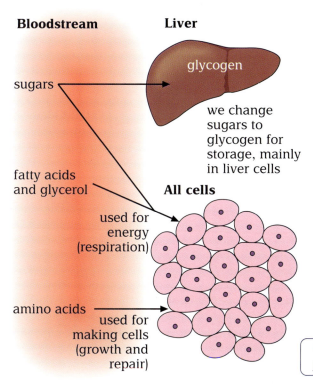

sugars

glycogen

we change sugars to glycogen for storage, mainly in liver cells

fatty acids and glycerol

All cells

used for energy (respiration)

amino acids

used for making cells (growth and repair)

What happens to the products of digestion

2 Describe <u>two</u> things that can happen to sugars after they pass into the blood.

3 Write down <u>one</u> kind of cell that uses lots of sugars. Explain your answer.

Remember: All the undigested food, including fibre, is got rid of in faeces. We say that we <u>egest</u> it. Faeces are mainly fibre, water and bacteria.

You should now have an understanding of these key ideas. You should also be able to spell and to know the meaning of the key words. The **key words** are in **bold** on this page.

To stay healthy we need a balanced diet.

The amount of each food group that we need depends on:
- our age
- our size
- our sex
- how active we are.

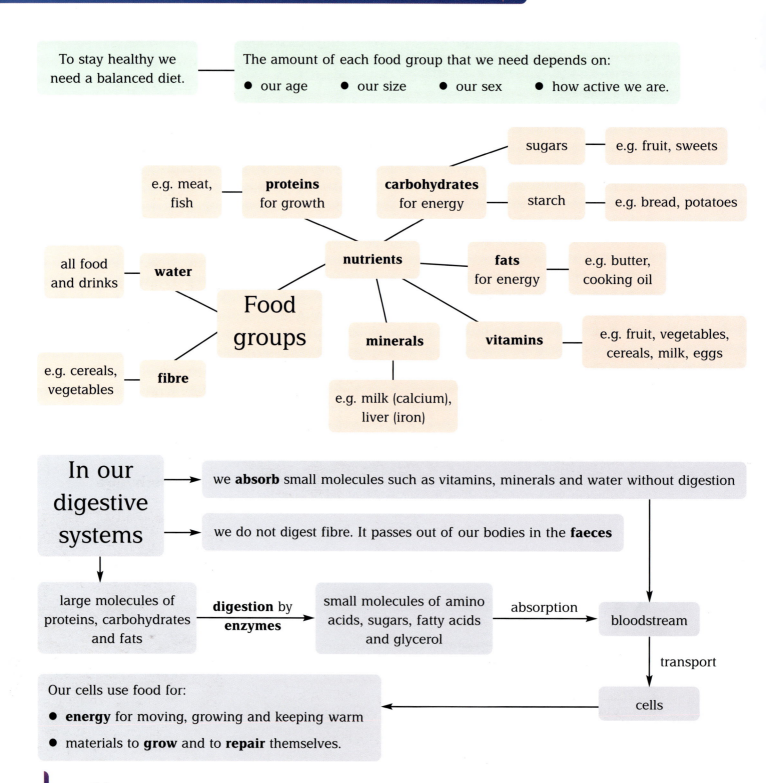

sugars — e.g. fruit, sweets

e.g. meat, fish — **proteins** for growth

carbohydrates for energy

starch — e.g. bread, potatoes

all food and drinks — **water**

nutrients

fats for energy — e.g. butter, cooking oil

Food groups

minerals

vitamins — e.g. fruit, vegetables, cereals, milk, eggs

e.g. cereals, vegetables — **fibre**

e.g. milk (calcium), liver (iron)

In our digestive systems

→ we **absorb** small molecules such as vitamins, minerals and water without digestion

→ we do not digest fibre. It passes out of our bodies in the **faeces**

large molecules of proteins, carbohydrates and fats — **digestion** by **enzymes** → small molecules of amino acids, sugars, fatty acids and glycerol — absorption → bloodstream

transport

cells

Our cells use food for:
- **energy** for moving, growing and keeping warm
- materials to **grow** and to **repair** themselves.

Respiration

To stay alive we must get energy from our food. In this unit, we shall be learning about food that our cells use as a source of energy, and how cells release the energy stored in this food.

KEY WORDS
glucose
amino acids
blood
moving
growing
keeping warm
energy
respiration
oxygen
aerobic
carbon dioxide
water
breathing
diffusion
heart
arteries
capillaries
veins
lungs
gas exchange

8B.1 How cells use food

You absorb digested foods such as **glucose** and **amino acids** into your **blood**. Your blood transports them to all the cells of your body. All cells use food as a source of materials to grow, and for energy. You use energy for **moving**, **growing** and **keeping warm**.

You need glucose for energy, and amino acids to make proteins for new cells.

your body needs new cells to repair damage

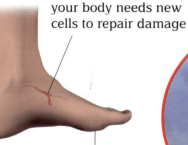

skin cells have to be replaced as they get worn away

Gail's muscle cells use up more glucose to release extra energy when she runs.

1   Write down <u>one</u> food that provides energy for cells.

2   Write down <u>two</u> reasons why you need to make new cells.

3   Write down <u>one</u> time when your muscle cells need more glucose than normal.

4   People doing sports often use high energy drinks. A slice of bread contains just as much energy as 200 cm³ of the drink.

Why is the energy in the drink more useful than the energy in the bread <u>during</u> exercise?

This is a high-energy drink. It contains a lot of glucose. High energy drinks are advertised for sports players.

How cells release energy from glucose

Glucose supplies your cells with **energy**. It is a type of sugar.

You could say that glucose is your body's fuel. In a machine like a car, fuel is burnt in oxygen to release the energy.

Your cells also use oxygen to release energy from their fuel. But the glucose doesn't burn. Chemical reactions in your cells break down the glucose and release the energy a bit at a time. We call this **respiration**. Respiration takes place in every cell in your body. Your cells normally use **oxygen** from the air when they respire. So we call this **aerobic** respiration.

Releasing energy from fuels

This is the word equation for respiration:

glucose + oxygen → **carbon dioxide** + **water** + energy

The word equation doesn't show that the glucose breaks down a bit at a time. It just shows the reactants and the products.

5 What is respiration?

6 Write down:

   a <u>two</u> things that cells need for respiration;

   b how these things get to your cells;

   c <u>two</u> waste substances that are produced in respiration.

8B.2 How oxygen reaches your tissues

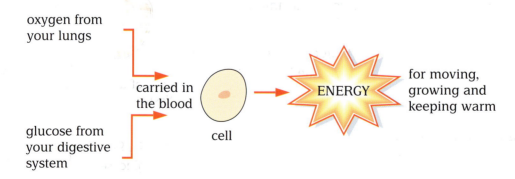

1 Why do we take oxygen into our bodies?

All your cells use oxygen to release energy from glucose.
Air contains oxygen. You take air in and out of your lungs.
This is called **breathing**. Some of the oxygen from the air in your
lungs passes into your blood. Your blood then carries the oxygen
to your tissues.

Oxygen and glucose pass out of the blood into the tissue fluid
(a liquid that surrounds all cells). Then they pass into the cells.
Cells use the oxygen to release energy from the glucose. The more
energy your cells use, the more oxygen and glucose they need.

2   Draw a flow diagram to show how oxygen gets from the
       air to the cells in the body.

3   Why do all the tissues in the body need blood vessels
       near them?

4   Some parts of the body have a better blood supply than
       others. Explain why the following organs need to have
       plenty of blood vessels:

       a the muscles;

       b the lining of the uterus of a pregnant woman.

More about exchanges

Substances are passing in and out of your blood and
your cells all the time. In Unit 7G, you learnt how
substances <u>diffuse</u> from where they are in <u>high</u>
concentrations to where they are in <u>low</u> concentrations.
Substances pass in and out of cells by **diffusion**.

In your tissues, oxygen and glucose diffuse into your
cells; carbon dioxide diffuses out of your cells into the
tissue fluid.

In your lungs, oxygen goes into your blood and carbon
dioxide leaves it and goes into the air in the lungs. When
you breathe out, you get rid of this extra carbon dioxide.

5   Write down <u>two</u> materials that diffuse:

       a from your blood to your cells.

       b from your cells into your blood.

6   Why do you need tissue fluid?

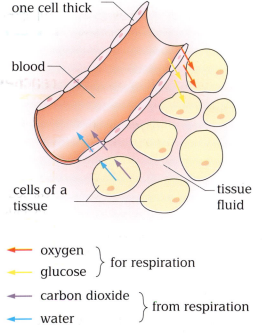

*Exchange of materials between
cells and the blood*

capillary wall
one cell thick

blood

cells of a
tissue

tissue
fluid

→ oxygen ⎫ for respiration
→ glucose ⎭

← carbon dioxide ⎫ from respiration
← water ⎭

Your heart

Your **heart** is a muscular pump that squeezes blood to move it around your body. It is divided down the middle by a wall of muscle. The left side pumps blood to the whole body. At the same time, the right side pumps blood to the lungs. Because of this, we say that it acts like a double pump.

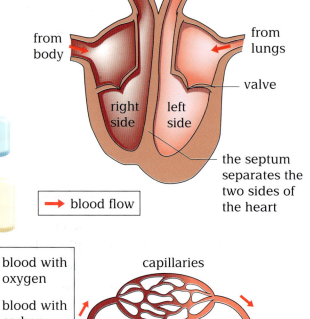

The human heart

to lungs   to body

from body

from lungs

valve

right side   left side

the septum separates the two sides of the heart

→ blood flow

7   What is your heart mainly made from?

8   Why does your heart need lots of glucose and oxygen?

The right side of your heart pumps blood to the lungs to pick up oxygen. The blood then goes back to the left side of your heart. The left side pumps blood rich in oxygen to the rest of your body.

9   Look at the diagram. Write down, in order, the parts that the blood goes through. Start and finish at the right side of the heart.

10   How many times does blood pass through your heart each time it does a full circuit of your body?

11   Why do you think the wall of the left side of the heart is thicker than the right side?

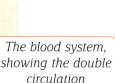

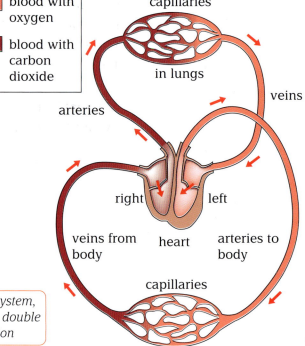

blood with oxygen

blood with carbon dioxide

capillaries in lungs

arteries

veins

right   left

veins from body   heart   arteries to body

capillaries

The blood system, showing the double circulation

Blood leaves your heart in **arteries**. Arteries split up into tiny tubes called **capillaries** in your tissues. This is where substances go in and out of your blood. Capillaries join up to form **veins** that take blood back to your heart.

12   Write down <u>one</u> reason why substances pass in and out of capillaries easily.

13   There are blood capillaries close to all the cells in all your organs, including your lungs. Why is this?

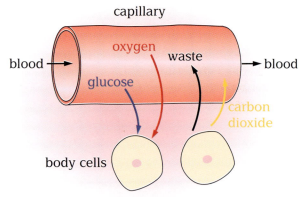

capillary

blood →   oxygen   waste   → blood

glucose   carbon dioxide

body cells

No cell is far from a capillary.
Capillary walls are only one wall thick.

Exchanging substances at the capillaries

The story of blood circulation

In Ancient Greek and Roman times, people thought that blood was one of four liquids that made up the body, and that the heart was where our emotions came from. They knew that blood moved out of the heart in blood vessels, but they thought the blood moved back and forth like the tides in the sea. They also believed that the body used up blood and that fresh blood was made from food, drink and air.

From Ancient times up to the Middle Ages, some of the knowledge of the body and how it worked was based on dissection – cutting up bodies to examine the parts. People didn't know that the heart was a pump or that blood was used for transport.

In the 17th century, scientists were trying to find things out by doing experiments. William Harvey was a British scientist who discovered that blood circulates around the body.
He observed the hearts of many different types of animals.
He compared the working of hearts and pumps. He also took measurements of the amount of blood leaving the heart.

William Harvey is often credited with the discovery of circulation. He showed that blood flows from the heart in arteries and back to the heart in veins.

14 How was the way that Harvey investigated the heart different from what was done in Ancient Greece and Rome?

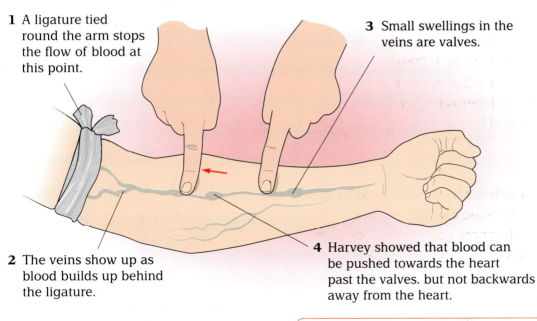

1 A ligature tied round the arm stops the flow of blood at this point.

3 Small swellings in the veins are valves.

2 The veins show up as blood builds up behind the ligature.

4 Harvey showed that blood can be pushed towards the heart past the valves. but not backwards away from the heart.

Harvey's experiment showed that blood circulates round the body in one direction.

Although Harvey worked out how blood moves around the body, he was not able to see the tiny blood vessels that link the arteries and veins together. We call them capillaries. They were discovered some years later by an Italian scientist called Marcello Malpighi. Capillaries are about 0.01 mm wide.

Marcello Malpighi found the capillary connection between arteries and veins that completes the circulation.

15 Why could Harvey not identify capillaries?

16 What technology was needed before capillaries could be discovered?

17 Draw a simple diagram to show what Harvey and Malpighi found.

18 What have modern scientists learnt from Harvey about studying the human body?

19 A lot more people were involved in the story of blood circulation than just Harvey and Malpighi.

  a Find out about the contribution made by one of the following to the story of blood circulation:

  • The Chinese

  • The Ancient Greeks (including scientists like Galen and Erasistratus)

  • Islamic science (including scientists like Ibn-al-Nafis).

  b Explain how our ideas of blood circulation have changed.

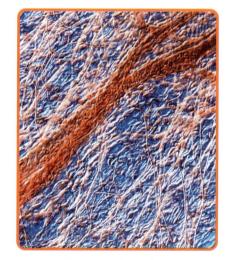

This photo of capillaries was taken using a modern microscope.

8B.3 What happens to oxygen when it reaches the cells

Aerobic respiration is a chemical reaction. It happens in every cell.

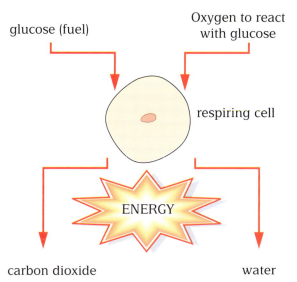

glucose (fuel)

Oxygen to react with glucose

respiring cell

ENERGY

carbon dioxide

water

1 Write down the word equation for respiration to remind yourself of what happens.

2 How do you think the amount of energy available is affected if there isn't enough:

a glucose?

b oxygen?

An aerobics exercise class

3 Look at the photograph. Are the people using more or less energy in the aerobics class than when they are walking along the street? Explain your answer.

4 Why do you think this exercise class is called 'aerobics'?

Sometimes there isn't enough oxygen in the air

This climber is working at 4500 m above sea level. At high altitudes like this the air molecules are more spread out. The climber is taking in less oxygen than normal with each breath. So his blood cannot supply his cells with all the oxygen they need.

5 What are the symptoms of altitude sickness?

6 The climber's cells are not getting enough oxygen. Explain why this makes him feel tired.

At high altitudes, the air in a passenger aeroplane is kept pressurised. This means that the amount of oxygen in the air is similar to that near the ground.

7 If the amount of oxygen in the air drops slightly, pilots notice that their judgement and ability to concentrate is not as good. Why do you think this is?

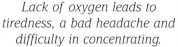

Lack of oxygen leads to tiredness, a bad headache and difficulty in concentrating.

8B.4 What happens in your lungs

You get the oxygen you need from the air. You breathe air in and out of your **lungs**. In your lungs, oxygen from the air diffuses into your blood. At the same time, waste carbon dioxide passes from your blood into the air. We call this **gas exchange**.

The air that you breathe out contains the waste carbon dioxide from respiration. So there is <u>less</u> oxygen and <u>more</u> carbon dioxide in the air that you breathe <u>out</u> than in the air that you breathe <u>in</u>.

1   Look at the diagram. Draw a flow chart to show the route air takes to get into your lungs.

2   What happens during gas exchange?

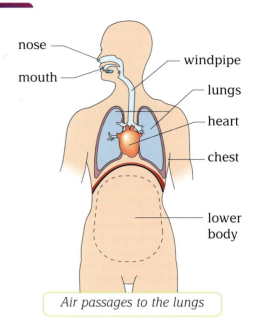

Air passages to the lungs

How gas exchange happens

Inside the lungs are millions of tiny air sacs called alveoli. Alveoli give the lungs a spongy feel and a very large surface area.

The walls of alveoli are only one cell thick. The very large number of capillaries surrounding the alveoli give the lungs their pink colour. Capillaries also have walls just one cell thick. So gases can pass quickly between the air in the alveoli and the blood.

Blood in the capillaries carries oxygen away continuously from the alveoli to supply the body cells. It also continuously brings waste carbon dioxide from the body cells to the alveoli. This prevents carbon dioxide building up in your blood and poisoning you. You breathe out the carbon dioxide.

3   Write down <u>two</u> reasons why it is important that gas exchange happens very quickly.

4   In what way are lungs arranged so that they have a large surface area?

5   Why does a large surface area help gas exchange happen quickly?

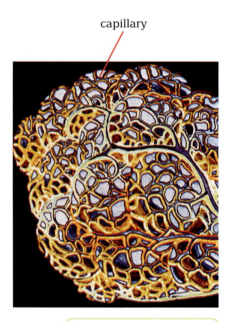

Alveoli in the lungs

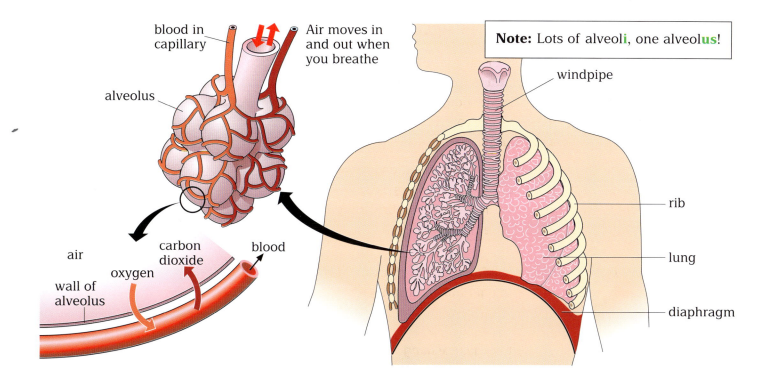

blood in capillary

Air moves in and out when you breathe

Note: Lots of alveol**i**, one alveol**us**!

alveolus

windpipe

rib

lung

diaphragm

air

carbon dioxide

oxygen

blood

wall of alveolus

6 Why do thin walls help gas exchange happen quickly?

7 Why can substances pass in and out of capillaries easily?

8 Why does having a lot of capillaries around the alveoli help gas exchange?

Smoking affects gas exchange

Smokers often cough because tobacco smoke irritates their breathing system. Coughing damages alveoli. The more they are damaged, the smaller the surface area of the lungs. This makes gas exchange more difficult.

healthy alveoli

damaged alveoli

9 How does the loss of alveoli affect the rate of gas exchange? Explain your answer.

10 Describe <u>one</u> feature of alveoli that makes them suitable for gas exchange.

8B.5 Comparing inhaled and exhaled air

Respiration makes waste products.

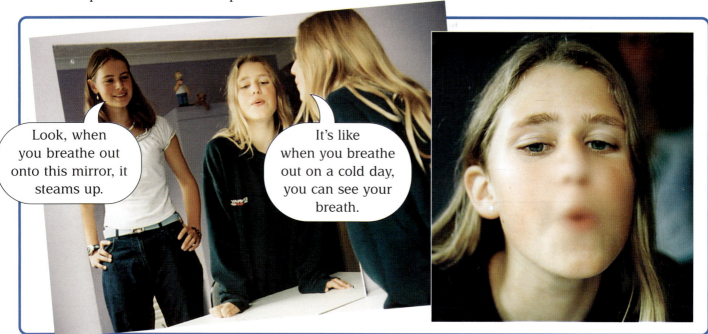

Look, when you breathe out onto this mirror, it steams up.

It's like when you breathe out on a cold day, you can see your breath.

1 What is the liquid on the mirror?

2 Where does it come from?

3 Why does water vapour show up in exhaled air on a cold day?

All cells make waste products when they respire. The waste products of aerobic respiration are carbon dioxide and water. Carbon dioxide is poisonous so you get rid of it in the air you exhale (breathe out).

Comparing the amounts of carbon dioxide in inhaled and exhaled air

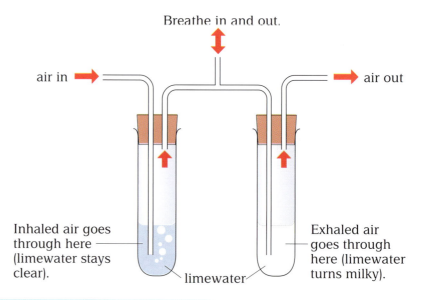

Breathe in and out.

air in

air out

Inhaled air goes through here (limewater stays clear).

Exhaled air goes through here (limewater turns milky).

limewater

4 What are the <u>two</u> waste products from aerobic respiration?

Peter did an experiment to compare the amounts of gases in air breathed in and out. Look at his results.

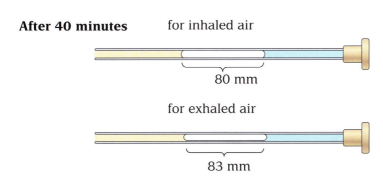

alkaline pyrogallate (absorbs oxygen)   original volume of air minus CO_2

At the start

screw for drawing up liquid

100 mm

After 40 minutes   for inhaled air

80 mm

for exhaled air

83 mm

5   **a** Which contains more oxygen – inhaled or exhaled air?

   b Explain why this is.

Gas	Air breathed in (%)	Air breathed out (%)
Oxygen	21	17
Carbon dioxide	0.03%	4
Nitrogen	79	79
Water vapour	Varies	Saturated

6   Which gas is more abundant in exhaled air than inhaled air?

7   Where was this extra gas made?

This athlete is measuring the gases in his exhaled air.

8   The amount of water vapour in inhaled air varies. Why is this?

9   The runner in the photograph is making more carbon dioxide than he does when he is asleep. Explain why that is.

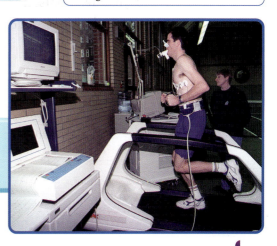

8B.6 Other living things respire too

The cells of <u>all</u> living things need to release energy to carry out their life processes. Most of them use oxygen and produce carbon dioxide. Carbon dioxide production is a good way of finding out if respiration is happening.

When you breathe in and out through limewater, you find that the air you breathe out contains a lot more carbon dioxide than the air you breathe in. But you can't ask a seed or a woodlouse to breathe in and out! The diagrams show what you can do.

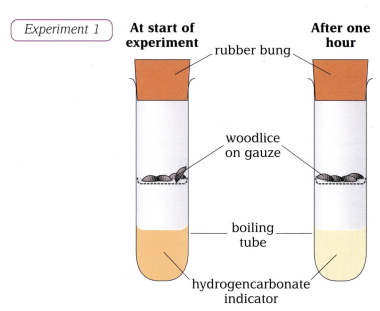

Experiment 1

At start of experiment — rubber bung — woodlice on gauze — boiling tube — hydrogencarbonate indicator

After one hour

Experiment 2

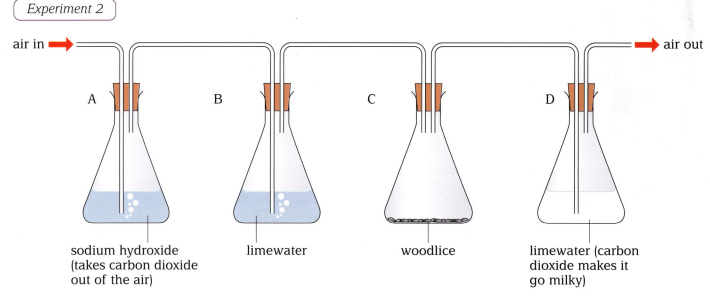

air in

A — sodium hydroxide (takes carbon dioxide out of the air)

B — limewater

C — woodlice

D — limewater (carbon dioxide makes it go milky)

air out

1 **a** Write down <u>two</u> substances that you can use to detect carbon dioxide.

   b Describe the change to each substance when carbon dioxide is present.

2 In experiment 2, the woodlice breathe in air that has no carbon dioxide in it. How do you know?

3 The change in the limewater shows that there is carbon dioxide in the air going through flask D.
Where did it come from?

In experiment 2, the first flask of limewater (flask B) shows that the air reaching the woodlice has no carbon dioxide in it.
We need flask B, to show that the carbon dioxide in flask D can have come only from the woodlice and anything living in or on them.

In experiment 1, we use one tube with and one tube without woodlice. Then we can say that the woodlice cause any change. We call the second tube the <u>control</u>. Without the control, we can argue that something else might have caused the change. Carbon dioxide could have leaked into the tube. Light or temperature differences, or anything else, could have caused the change.

4 Describe the results of experiment 1.

5 What can you conclude from experiment 1?

6 Katie did an experiment like this with maggots. But she didn't use a second tube. The indicator changed colour. Her teacher told her that she couldn't conclude that maggots produced carbon dioxide. Why is this?

When you design experiments using living things, remember to:

- use more than one living thing, because living things vary;

- use a control, then any change can be caused only by the variable that you are testing – it can't be caused by light, or temperature, or any other variable;

- vary only one thing at a time.

Remember that all living things respire.

| lion | dandelion | oak tree | germinating seed | bacterial cell | killer whale | yeast cell |

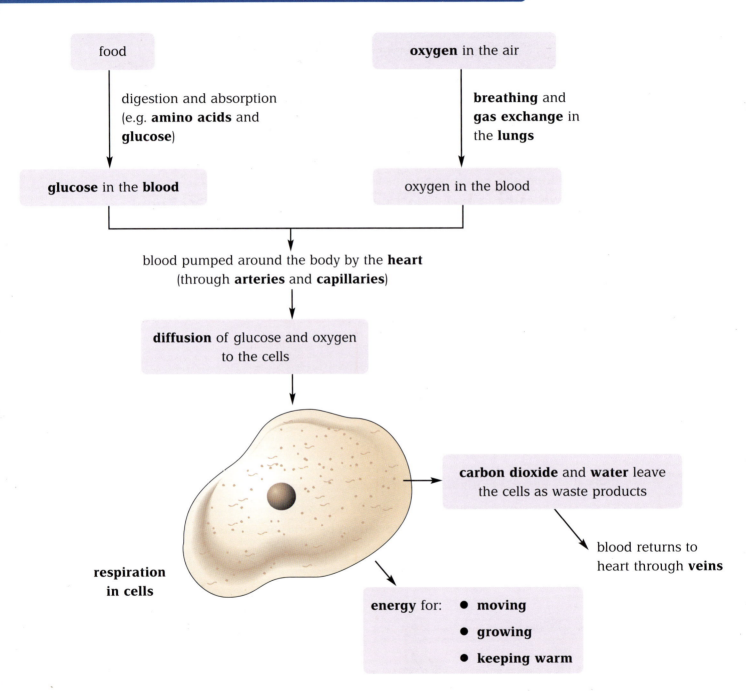

8B

You should now understand the key words and ideas shown below.

food

oxygen in the air

digestion and absorption (e.g. **amino acids** and **glucose**)

breathing and **gas exchange** in the **lungs**

glucose in the **blood**

oxygen in the blood

blood pumped around the body by the **heart** (through **arteries** and **capillaries**)

diffusion of glucose and oxygen to the cells

carbon dioxide and **water** leave the cells as waste products

blood returns to heart through **veins**

respiration in cells

energy for:
- **moving**
- **growing**
- **keeping warm**

The word equation for **aerobic** respiration is:

glucose + oxygen → carbon dioxide + water + energy

Microbes and disease

In this unit we shall be learning about some micro-organisms and how we grow them to make useful products. We shall also find out about micro-organisms that cause disease and how our bodies fight disease.

KEY WORDS
micro-organisms
viruses
bacteria
fungi
disease
infection
immunity
antibiotic
vaccine
immunisation

8C.1 Micro-organisms and how to grow them

Types of micro-organisms

Some living things are so small that we can only see them through a microscope; we call these tiny living things **micro-organisms** or microbes. They include **viruses** and **bacteria** and some **fungi**.

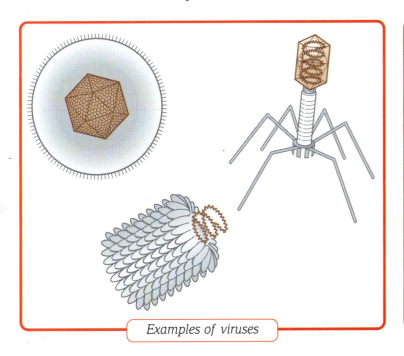

Examples of viruses

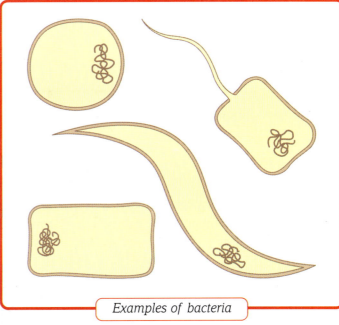

Examples of bacteria

There are lots of different types of micro-organisms. Although they are very small, they have a huge effect on our lives.
Micro-organisms are an essential part of life on Earth and do many vital jobs. They also cause many problems, including **disease**.

1   Write down <u>three</u> different types of micro-organism.

Here are two micro-organism fact files:

FACT FILE: Viruses

Average size:	0.0001mm
Structure:	A strand of genetic material wrapped in a protein coat.
Found:	Viruses can reproduce only inside living cells.
Uses:	To kill pest animals.
Diseases:	Common cold, influenza (flu), measles, AIDS, yellow fever, rabies. Viruses cause disease in animals, plants and even other micro-organisms.

FACT FILE: Bacteria

Average size:	0.001mm
Structure:	Bacteria are single-celled with a strong cell wall. Their genetic material is not in a nucleus.
Found:	Most bacteria live in water, soil and decaying matter.
Uses:	To make yoghurt, cheese and vinegar.
Diseases:	Typhoid, cholera, food poisoning.

2 How many times bigger is the average bacterium than the average virus?

Some fungi are micro-organisms, for example yeast and mould. Yeast is made of single cells that reproduce by budding off new cells. Yeast cells are larger than viruses.

Moulds are fine threads that grow on rotting food; they also give blue cheese its colour and flavour.

Fungi can be helpful; we use yeast to make bread, wine and beer and mould to make antibiotics. They can also be harmful; fungi cause athlete's foot and ringworm.

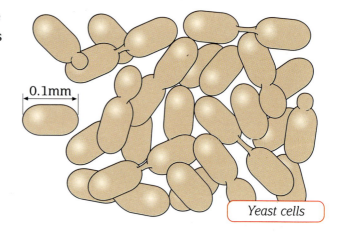

0.1mm

Yeast cells

3 Which type of micro-organism is the smallest?

4 Which <u>two</u> types of micro-organism make food rot?

5 Use the headings of the fact files to make a table of information about viruses, bacteria and fungi.

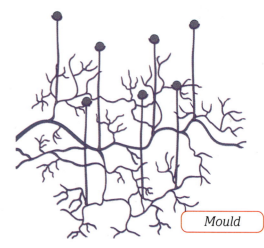

Mould

How to grow micro-organisms

Yeast is the most commonly grown micro-organism in the world. We use it to make beer, wine and bread. Yeast is a living thing, so it needs warmth and food to grow.

Just like all living things, yeast breaks down food to get energy; we call this process respiration. If it uses oxygen, we call it aerobic respiration. But yeast can also respire without oxygen. When it does this it makes alcohol. In both cases, it produces carbon dioxide.

6 Write down <u>three</u> things that both yeast and humans do.

7 The yeast cells in this vat respire aerobically. How can you tell this from the design of the vat?

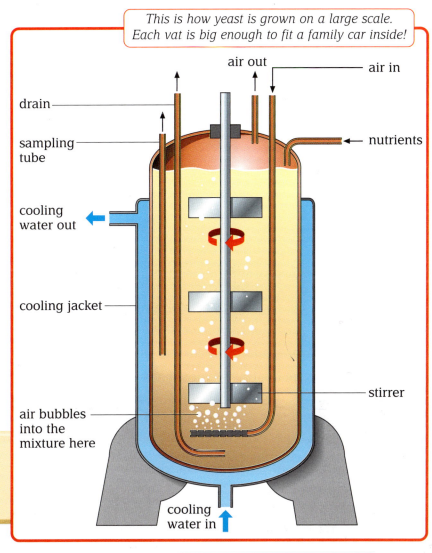

This is how yeast is grown on a large scale. Each vat is big enough to fit a family car inside!

air out · air in

drain

sampling tube

nutrients

cooling water out

cooling jacket

stirrer

air bubbles into the mixture here

cooling water in

Bread dough is made of yeast, flour, water and a little sugar and salt. Laura did an experiment to see if sugar helps dough rise. She measured the amount of dough in each measuring cylinder after 30 minutes.

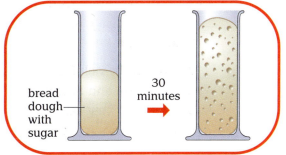

bread dough with sugar

30 minutes

8 How does the amount of sugar affect how high the dough rises?

9 What is the gas that makes dough rise?

10 When making bread, you leave the dough to rise for several hours. Then you bake it for 30 minutes at 200°C. What do you think happens to yeast during baking? (Remember: yeast is a living thing.)

Amount of sugar in the dough (g)	Volume of the dough after 30 minutes (cm³)
0	40
5	52
10	66
15	74
20	80

Yeast is just one type of fungus. We can also grow other fungi in large vats to make useful products.

Product made by fungi	Use
Penicillin	An antibiotic to treat some diseases
Citric acid	Added to make soft drinks taste tangy
Cortisone	To treat arthritis
Pectinase	Added to fruit juice to make it clear
Mycoprotein	A substitute for meat that is suitable for vegetarians (for example, Quorn)

 11 Look at the table. Write down <u>two</u> medicines made by fungi.

Growing micro-organisms in a laboratory

In a laboratory we grow bacteria in Petri dishes – small plastic or glass dishes with lids. Food for the bacteria is mixed with a jelly called agar, which is made from seaweed. Each type of micro-organism needs its own particular balance of minerals and food to grow.

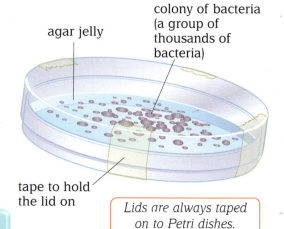

agar jelly

colony of bacteria (a group of thousands of bacteria)

tape to hold the lid on

Lids are always taped on to Petri dishes.

 12 Bacteria are so small that you can only see them with a microscope. Why can you see bacteria growing on agar?

 13 Why do you think the dishes used to grow bacteria are also called agar plates?

We must clean everything we use to grow micro-organisms before it is used. Petri dishes and agar are heated to over 100 °C to sterilise them. It is important to keep all benches and equipment clean.

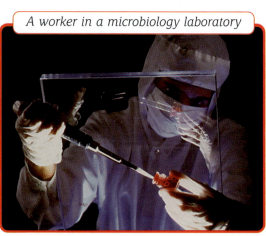

A worker in a microbiology laboratory

 14 Look at the photograph. Write down <u>three</u> things the worker is wearing to prevent any unwanted micro-organisms from contaminating his work.

 15 What do you think would happen if laboratory workers did not sterilise the equipment properly?

8C.2 Micro-organisms and disease

Illnesses caused by micro-organisms are called **infections**, or infectious diseases. They can be passed from one person to another. Some are spread in contaminated food and water.

Diseases can spread:

- by droplet, e.g. tuberculosis (TB) bacteria and chickenpox virus;

- by animals, e.g. malaria and yellow fever by mosquitoes, and rabies by mammals;

- in food, e.g. *Salmonella* bacteria from flies, dirty hands, or dirty knives and dishes;

- in contaminated water, e.g. typhoid bacteria.

1 How can a child at a birthday party give all the other children chickenpox?

2 Why is it important that you wash your hands before you prepare food?

3 How could you stop food poisoning being spread by flies?

4 Write down <u>one</u> way of making water safe to drink.

How human-to-human contact spreads diseases

Athlete's foot is a fungal disease spread by touch. Some diseases such as AIDS are spread when people have unprotected sex.
A pregnant woman can pass on a disease to her unborn child.
A baby can also catch a disease from his or her mother through breast milk.

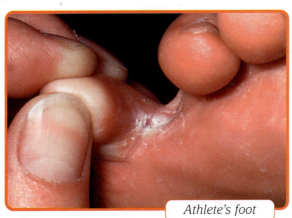

Athlete's foot

5 Write down <u>two</u> ways that a mother can pass on a disease to her baby.

How to stop an epidemic

An epidemic is an outbreak of a disease affecting a large number of people in a population.

In 1854 there was a terrible epidemic of a disease called cholera in London. In just a few weeks, thousands of people caught cholera in a small area of London called Soho. Over 600 people died. Cholera was a common disease in polluted city areas, so people thought that you caught cholera from bad air.

This is how quickly the disease spread in the first three days of the epidemic:

Date in 1854	Number of new cases of cholera	Number of deaths from cholera
31st August	56	3
1st September	143	70
2nd September	116	127

6   How many people died from cholera in the first three days of the epidemic?

Many more people would have died had it not been for a doctor called John Snow. He was convinced that cholera was caused by infected water rather than bad air. He heard about the deaths in Soho and went to investigate.

At that time few people had running water in their houses and mostly had to rely on public street pumps for all their water. John Snow suspected that people became infected when they drank the water from the Broad Street pump in Soho.

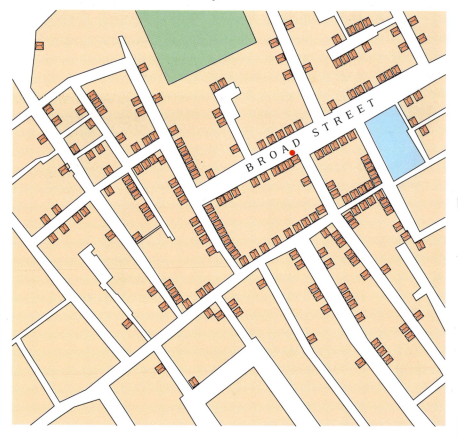

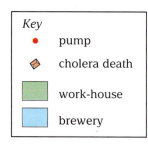

Key
- • pump
- ◇ cholera death
- ▢ work-house
- ▢ brewery

This was John Snow's evidence:

- Most people who died lived very near to the Broad Street pump.

- Only 5 out of 500 people died at the workhouse round the corner from the pump. The workhouse had its own well.

- No one in the local brewery died. All the workers drank beer instead of water.

- Two ladies who lived 8 km away died of cholera. They had a bottle of water brought to them from the Broad Street pump because they liked the taste.

By 7th September three-quarters of the people living in Soho had fled the area. Many of those who remained were ill and 28 more died that day. John Snow put his evidence to the Parish Board and they agreed to remove the handle of the pump the next day. People could no longer drink the water. The number of new cases began to fall.

7 Imagine you are the person who took the handle off the pump. A crowd of local people complain to you that now they will have to walk 10 minutes to get drinking water. What do you say to them?

A few months later John Snow found the cause of the epidemic.

- Water for the pump came from an underground well.

- Number 40 Broad Street had an underground cesspit for sewage.

- During August a baby at number 40 was ill with cholera and his mother washed his nappies in water that she then tipped into the cesspit.

- The cesspit wall was cracked and the sewage leaked out into the nearby well.

- This allowed the bacteria that cause cholera to reach the Broad Street pump.

8 John Snow could not see the bacteria that cause cholera. How could he be sure there were bacteria in the well of the Broad Street pump?

8C.3 Protecting ourselves against disease

Your body defends itself against micro-organisms in several ways.

1 Why can't micro-organisms normally get into your body through your skin?

2 In which part of your body are micro-organisms killed by acid?

3 Animals often lick their wounds. How does this help them to heal?

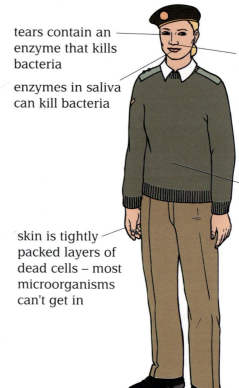

tears contain an enzyme that kills bacteria

enzymes in saliva can kill bacteria

hairs in your nostrils trap microorganisms and mucus contains enzymes that can kill bacteria

skin is tightly packed layers of dead cells – most microorganisms can't get in

most bacteria swallowed with food and drink are killed by acid in the stomach

But micro-organisms get past these defences!

Tuberculosis (TB) is a serious disease that destroys lung tissue. When someone has TB they have a bad cough and each time they cough they spray little droplets into the air. These droplets contain TB bacteria. If you are nearby, you can breathe them in. Sometimes bacteria get past the defences in your air passages and into your lungs.

4 Why are you less likely to catch TB if you breathe through your nose rather than through your mouth?

What happens when you catch a cold

Several different viruses cause colds. Most adults catch two or three colds a year; many children catch up to eight colds a year. When you have a cold and sneeze, you spray droplets of liquid into the air. Each droplet contains thousands of virus particles. When you first catch a cold, the virus multiplies very quickly; you start to feel ill, with a runny nose, sore throat and a cough. Then your body fights back against the invasion; cold viruses are destroyed and you start to feel better.

5 Why do you cover your nose and mouth when you sneeze?

6 Suggest <u>one</u> reason why children catch more colds than adults.

How your body destroys micro-organisms

Your blood contains red blood cells and white blood cells. There are two different types of white blood cells.

- One type of white blood cell engulfs (traps) micro-organisms and destroys them.

- Another type of white blood cell makes disease-fighting substances called antibodies. These can stop micro-organisms from causing disease.

Each antibody that you make is specific to fighting one type of micro-organism – an antibody won't work against any other type of micro-organism. Different micro-organisms need different antibodies. It takes time for your body to make these different antibodies, and you feel ill until you have made enough antibodies to destroy the micro-organisms.

Once you have had a disease, your white blood cells have learnt how to make the antibodies. They will be able to make the right ones much more quickly in future. If a second attack comes, your body can destroy the micro-organisms before they have time to make you ill. This means you are immune to the disease. You have **immunity**.

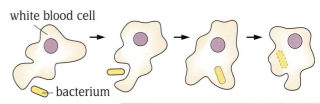

The micro-organism is taken in to the white blood cell and destroyed.

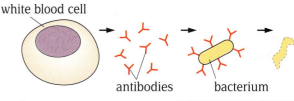

A white blood cell that produces antibodies.

7 Why don't you usually catch the same disease twice?

8 When you have had a cold you become immune to that particular virus. How is it that a few months later you can catch another cold?

9 Breast milk contains the mother's antibodies. How does this stop a breast-fed baby from catching some diseases?

Antibiotics

Antibiotics are substances made by one living thing that can kill another living thing. They were discovered by accident!

Mould is a type of fungus. For centuries, people knew that spreading mould on a wound or cut often helped it to heal. Scientists also noticed that mould could stop bacteria from growing, but they didn't realise how important this discovery was.

Mould often grows on rotten food. The mould on this orange is one species of Penicillium.

10 Would you put mouldy food on a cut to help it to heal? Explain your decision.

In 1928 some spores of mould landed by accident on a Petri dish of bacteria belonging to Alexander Fleming. He noticed that, where the mould grew, there were no bacteria. The mould had made a substance that stopped the bacteria growing. Since the mould was called *Penicillium*, Fleming named the substance penicillin.

11 What stopped bacteria from growing close to *Penicillium*?

In 1935, a team of Oxford scientists led by Howard Florey and Ernst Chain began to make and test penicillin. At first, they had only enough penicillin for tests on mice and a few patients. By 1942, during the Second World War, they could make large amounts. So penicillin was used to save sick and wounded soldiers.

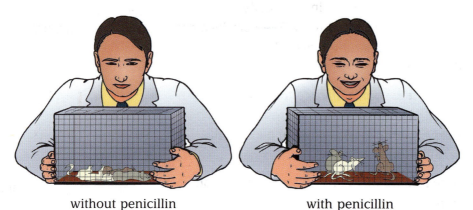

without penicillin   with penicillin

12 Normally new drugs are tested and trialled for years before they are widely used. Why was penicillin used as soon as it could be made in large quantities?

13 Before penicillin was discovered, 1 in 3 people who caught pneumonia died. When penicillin was used, only 1 in 20 people died. If 300 people caught pneumonia, how many would probably die:

   a before penicillin?   **b** after penicillin?

14 Explain what the discovery of penicillin tells us about the way scientific knowledge can develop.

Penicillin was the first antibiotic discovered and its use has saved millions of lives. Since then, scientists have extracted other important antibiotics from different moulds and soil bacteria. These antibiotics can kill some of the bacteria that penicillin could not.

15 Mrs Sharples has flu (caused by a virus). She wants antibiotics to make her better. What would you tell her?

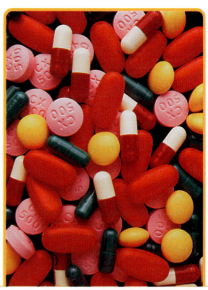

Different antibiotics treat different infections, but none of them has any effect on diseases caused by viruses.

Jabs

A jab is an injection; it contains a **vaccine** to make you immune to a particular disease. Your TB jab is just one of 11 injections you may have during your childhood. The **immunisation** for tuberculosis (TB) is called the BCG jab after two scientists who developed it: Calmette and Guérin.

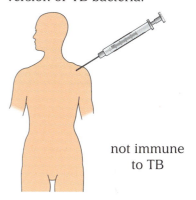

A TB jab contains a weak version of TB bacteria.

not immune to TB

16 Find out what vaccinations you have had, and when you had them.

How the TB jab works

The vaccine in a TB jab contains an extract of the bacteria that cause tuberculosis. The white blood cells in your body respond to the vaccine as if you've been infected with TB. They produce the correct antibodies to kill the bacteria that cause TB. Your arm may feel sore while this happens.

In future, your white blood cells will be able to make the right antibodies much more quickly. So if TB bacteria get into your body, you will make antibodies to destroy them and not become ill.

Before scientists developed the TB jab, the only way you could be immune to TB was by surviving the disease. A jab makes you immune without having the disease itself.

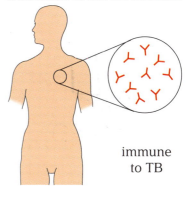

White blood cells make the right antibodies to kill TB bacteria.

immune to TB

17 Why do your white blood cells make antibodies in response to the TB vaccine?

18 Draw a flowchart to show how the TB immunisation stops you catching TB.

19 Look at the graph. What year do you think TB jabs were first introduced as a routine immunisation for all children in England and Wales?

20 In the year 2000, there were still over 6000 cases of TB in England and Wales. Not everyone is immunised against TB. Write down <u>two</u> possible reasons why someone may not have had the TB jab.

Total annual number of cases of TB in England and Wales, 1940 to 2000. (From the Public Health Laboratory Service)

You should now understand these key words and ideas.

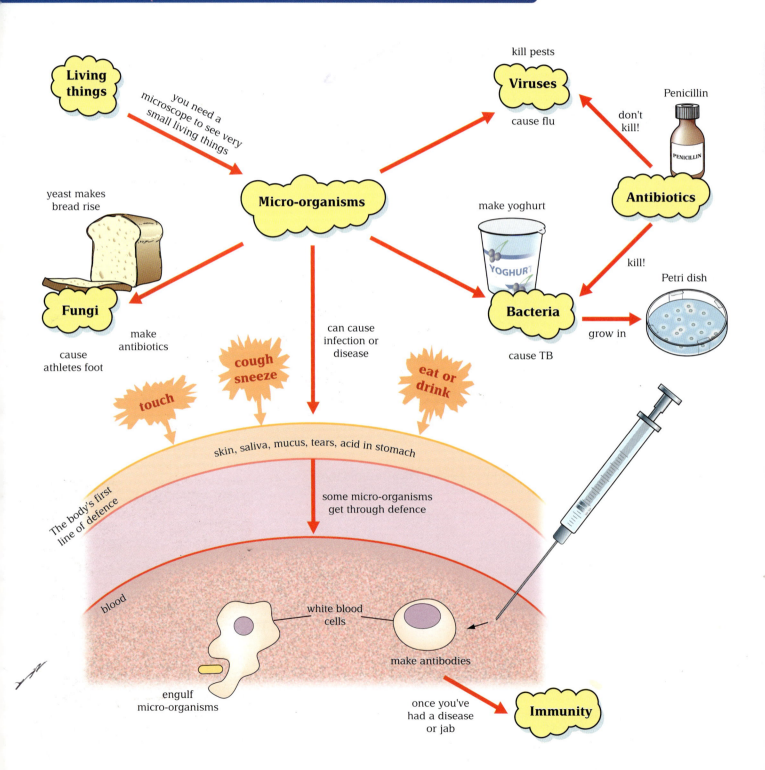

kill pests

Living things

you need a microscope to see very small living things

Viruses

cause flu

Penicillin

don't kill!

Antibiotics

yeast makes bread rise

Micro-organisms

make yoghurt

kill!

Fungi

make antibiotics

cause athletes foot

Bacteria

cause TB

grow in

Petri dish

can cause infection or disease

touch

cough sneeze

eat or drink

skin, saliva, mucus, tears, acid in stomach

The body's first line of defence

some micro-organisms get through defence

blood

white blood cells

make antibodies

engulf micro-organisms

once you've had a disease or jab

Immunity

Ecological relationships

In this unit we shall extend our study of classification to include plants. We shall then look at how plants and animals in a community affect each other, and how their environment affects them.

KEY WORDS
habitat
environmental
  conditions
adapted
vertebrates
invertebrates
community
population
quadrat
food chain
producer
consumer
food web
herbivore
carnivore
pyramid of numbers

8D.1 Animals, plants and adaptations

You learned in Unit 7C that the place where a plant or animal lives is called its **habitat** and that different habitats have different **environmental conditions** such as temperature and amount of water.

The habitat of a plant or animal must provide all the things that it needs to live and reproduce. Different plants and animals have different needs. For example a gerbil can survive on less water than a mouse. This is because it has the features that help it to survive. We say that it is **adapted** to its environment.

Different plants and animals are adapted to live and reproduce in different habitats.

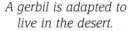
A gerbil is adapted to live in the desert.

1 What is the meaning of the word habitat?
2 Explain why organisms are adapted to their environments.

When you study a habitat you need to be able to identify the plants and animals that you find. It helps if you know which group they belong to. When we put plants and animals into groups we say that we <u>classify</u> them.

Classifying animals

In Unit 7D you learned how we classify animals.

- Animals with backbones are called **vertebrates**.

- Animals without backbones are called **invertebrates**.

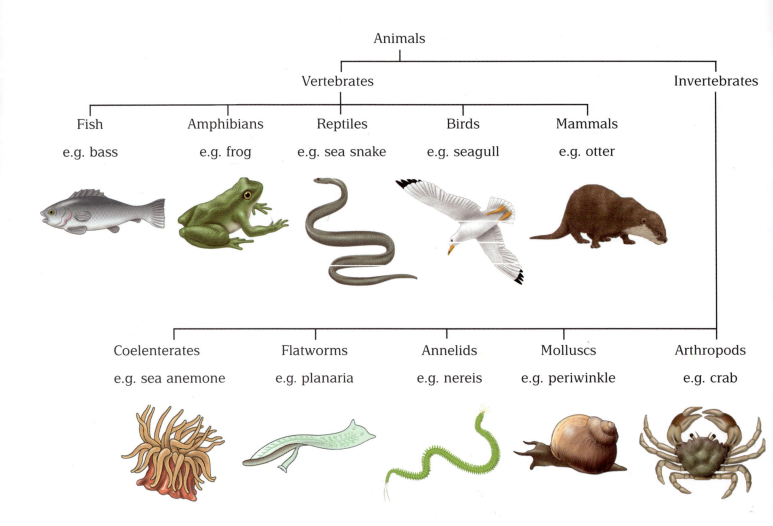

Animals

Vertebrates   Invertebrates

Fish   Amphibians   Reptiles   Birds   Mammals
e.g. bass   e.g. frog   e.g. sea snake   e.g. seagull   e.g. otter

Coelenterates   Flatworms   Annelids   Molluscs   Arthropods
e.g. sea anemone   e.g. planaria   e.g. nereis   e.g. periwinkle   e.g. crab

3   **a** Name the <u>two</u> main groups of animals.

    b Which of the two groups has an inside skeleton?

4   **a** Write down the <u>five</u> groups of vertebrates.

    b For each group, write down <u>one</u> feature that makes it different from the other groups.

5   Name some groups that invertebrates are divided into.

Now let's classify green plants

There are hundreds of thousand of different plants, so we divide them into smaller groups to help us to identify and to study them. There is more than one way of doing this.

Some plants have a special transport system for food and water called a vascular system. So we call these plants <u>vascular plants</u>.

Plants without a vascular system are called <u>non-vascular plants</u>.

xylem carries water and salt

phloem carries sugars

A slice through the root of a vascular plant

Vascular and non-vascular plants are divided into smaller groups.

6 Name <u>two</u> types of vascular tissue found in vascular plants.

7 Why do we divide plants into groups?

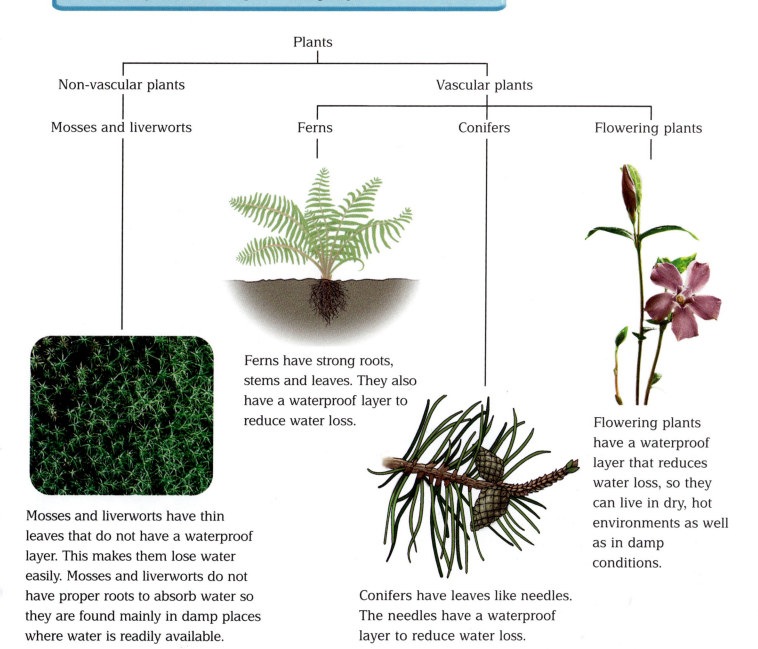

Plants

Non-vascular plants

Mosses and liverworts

Vascular plants

Ferns

Conifers

Flowering plants

Ferns have strong roots, stems and leaves. They also have a waterproof layer to reduce water loss.

Mosses and liverworts have thin leaves that do not have a waterproof layer. This makes them lose water easily. Mosses and liverworts do not have proper roots to absorb water so they are found mainly in damp places where water is readily available.

Conifers have leaves like needles. The needles have a waterproof layer to reduce water loss.

Flowering plants have a waterproof layer that reduces water loss, so they can live in dry, hot environments as well as in damp conditions.

8 Explain why mosses and liverworts often live in a damp environment.

9 Why are vascular plants able to live in a wide range of habitats?

8D.2 Interactions in a habitat

In this topic you will be looking at some ways of finding out about habitats and the plants and animals that live in them.

Collecting data to answer questions about a habitat

Molly and her class did some fieldwork on a rocky shore. This is an interesting place to carry out work because you can find lots of plants and animals here. Animals are harder to see in many other habitats.

When the tide comes in, it brings with it a supply of food for the animals. The food includes a range of microscopic plants and animals floating in the water that together are called <u>plankton</u>. These are the food for many animals living on the shore.

The plants and animals found on the shore depend on each other. They are called a **community**.

When the tide comes in it brings with it tiny plants and animals called plankton.

> **1** Write down <u>one</u> advantage and <u>one</u> problem of doing fieldwork on a rocky shore.

It is easier to study the shore when the tide is out, but the seaweed makes it slippery.

Asking questions before the visit

In the lesson before the visit, Molly's teacher asked the class to think about what they'd like to find out about the rocky shore. Molly and her friend came up with these ideas.

> What plants and animals live on the rocky shore?
>
> Will we find the same plants and animals on all parts of the shore?
>
> How can we find the numbers of the different plants and animals?
>
> Will different areas of the shore have different environmental conditions?

2 Molly's friend David said, 'We can just count all the plants and animals that we see.' Write down why this is not a good idea.

Often it is not possible to count all the individuals in a **population** so we take a <u>sample</u>. Then we estimate the number of organisms that live in an area.

3 Write down <u>two</u> safety points that the class needs to think about before working on the rocky shore.

Collecting the information

Molly and her group decided to see if different plants and animals lived on different parts of the shore.

Molly's teacher gave her a **quadrat** to help her to do this. She threw a plastic card on the ground and placed the quadrat so the card was in the centre. Molly then wrote down the names of the different plants and animals that were in the quadrat.

She did this 10 times near the top of the shore. She then moved nearer the sea and used the quadrat another 10 times. When she completed this, she used the quadrat another 10 times in the middle of the shore.

Molly's teacher thought that this was a good idea and asked David's group to do the same experiment. The teacher said that they would compare the results when they got back to school.

Name	Quadrat number				
	1	2	3		
Periwinkle	III	II	₩₩₩		
Anemone	₩₩₩	III	I		
Barnacle	IIII	I			

4 Why was it a good idea for several groups to do the same experiment?

Finding out what lives on the shore

When Molly looked at her results she could see that different plants and animals lived in different areas of the shore.

Molly's teacher asked her to call the area nearest the sea the <u>lower shore</u>, and the area furthest from the sea the <u>upper shore</u>. The area between the two is called the <u>middle shore</u>.

The upper shore spends a lot of time not covered by water. It is exposed to different weather conditions.

The middle shore spends less time under water than the lower shore. It is not exposed to the weather for as long as the upper shore.

The lower shore spends a lot of time under the water.

Molly's teacher explained that the different areas of the shore are examples of different habitats. Different habitats support different living things.

5   Which of these areas is covered by the sea for the longest time?

Molly summarised her results in a table.

Area	Species found
Upper shore	Most of the rock is covered by black lichen. In rock crevices I found a few tiny periwinkles and barnacles.
Middle shore	I found lots of barnacles, limpets and mussels in this area of the shore. I also found some small pieces of seaweed in this area.
Lower shore	The seaweed was long and flexible. I could not remove it from the rock. I found crabs, starfish, limpets, fish and sea anemones here.

6   Look at the information in Molly's table. Write down which area of the shore has the widest variety of living organisms.

7   Molly's friend Patrick suggested that it would be a good idea to take some of the limpets back to the laboratory to look at in more detail. Why must they <u>not</u> do this?

Explaining the differences

When the class returned to school the teacher asked them to try to explain why there were different communities in different habitats. They thought that the environmental conditions were different in the different areas of the shore.

These are their ideas:

- **Upper shore** – This area can dry quickly because it spends a lot of the time not covered by the seawater. This means that there is less feeding time for the animals. It is often exposed to very hot or very cold weather conditions. Sometimes rain makes the water less salty.

- **Middle shore** – This area spends more time under water than the upper shore, but less time under water than the lower shore. The water brings with it a rich supply of food.

- **Lower shore** – This area spends most of the time under water. This means that there is less variation in temperature and not such a problem of drying.

Lichens are made of a green alga and a fungus. They often live on rocks and tree trunks. They can survive because they take a long time to dry out.

Many species of seaweed are long and flexible. They also have very strong holdfasts, which fix them to the rocks.

Limpet shells fit closely to the rock so that they don't dry out. They feed on tiny seaweeds on the rock when the tide is in.

Plants and animals on the shore have ways of making sure that the waves don't wash them away.

Mussels live attached to rocks. They filter plankton from the water when the tide is in.

Barnacles are cemented to the rock. When the tide is in they filter plankton from the water.

8 Explain why seaweeds have strong holdfasts.

9 Lichens take a long time to dry out. Why is this useful for an organism that lives on the upper shore?

Population size and environmental conditions

Molly's teacher agreed with their ideas. She said that the size of the population of an organism is affected by environmental conditions, such as amount of light, water and nutrients. The teacher also said that organisms will have more chance of survival where there is less variation in temperature.

Molly looked at the results of her quadrats for the population of barnacles.

10 Which area of the shore contained the largest population of barnacles?

Area	Number of barnacles
Upper shore	5
Middle shore	62
Lower shore	0

11 Molly said, 'More barnacles live on the middle shore than the upper shore because the barnacles on the middle shore have more time to feed from the water.'

Do you agree or disagree with Molly's conclusion? Explain your answer.

12 Molly made the conclusion just from looking at her own results. If she is a good scientist what should she do to make sure that the conclusion is correct?

13 Explain <u>one</u> other factor that makes it easier for barnacles to survive in the middle shore than the upper shore.

8D.3 How living things depend on each other

A **food chain** shows how energy is transferred from one organism to the next in a community.

Each food chain starts with a green plant, which is called a **producer**. An animal that feeds on green plants or other animals is called a **consumer**.

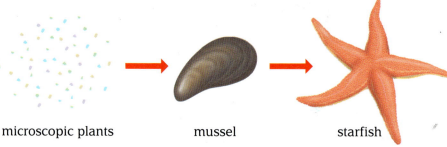

microscopic plants   mussel   starfish

A simple food chain

In most communities plants and animals belong to more than one food chain. A number of food chains joined together is called a **food web**. This gives a more complete picture of how an animal feeds.

The diagram shows part of a seashore food web.

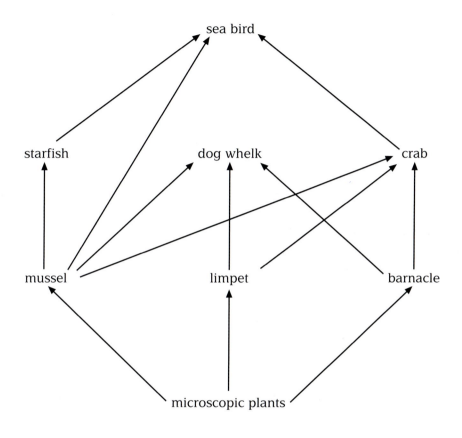

sea bird

crab

dog whelk

starfish

barnacle

limpet

mussel

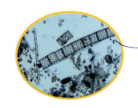
microscopic plants

1   From the food web, write down:

   a the producer;

   b <u>one</u> consumer.

2   Use the food web to find <u>two</u> things that crabs eat.

3   Write down <u>two</u> food chains in the food web that end with a sea bird.

An animal that only eats plants is called a **herbivore**. An animal that feeds on other animals is called a **carnivore**.

4   Name <u>one</u> herbivore in the food web.

5   Name <u>one</u> carnivore in the food web.

Using food webs

We can use food webs to predict the effect of a rise or fall in the population size of a particular plant or animal in a community.

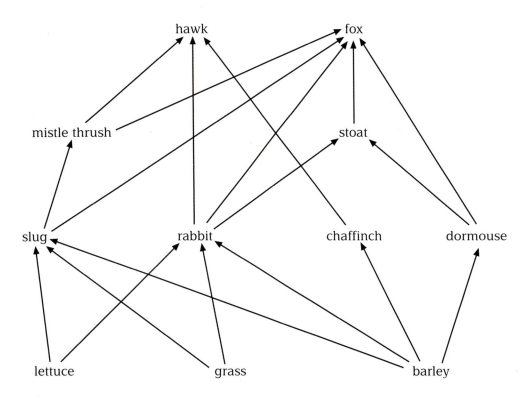

fox

hawk

stoat

mistle thrush

dormouse

chaffinch

If the population of one organism goes up or down it affects the rest of the food web.

If the rabbits are killed by a disease:

- the number of dormice may decrease – the stoats have fewer rabbits to eat, so they eat more dormice;

- the number of stoats may decrease because they have lost an important food source.

6 What will happen to the lettuce population if all the rabbits are killed?

7 What effect will the fall in rabbit population have on the slug population? Explain your answer.

lettuce

grass

barley

slug

rabbit

Pyramid of numbers

In a food chain, not all of the energy taken in by an organism passes to the next organism. Some is used for movement, growth and warmth. So there is less energy for the organisms at each stage.

Because of this, the number of animals gets lower as the food chain goes from stage to stage.

100 lettuces → 10 rabbits → 1 fox

We can show the change in population as we move along a food chain by a **pyramid of numbers**.

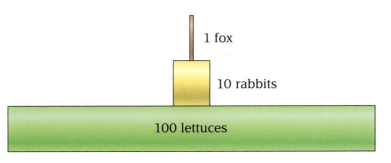

8 Draw the shape of the pyramid of numbers for the following food chains.

a dandelions → rabbits → fox

b microscopic plants → insect larvae → perch → pike

A problem with pyramids of numbers is that they do not allow for the size of the organism at each level of the food chain.

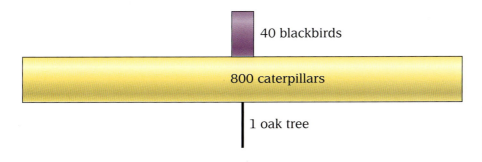

An oak tree is very big compared to a caterpillar. There is only one oak tree, but it has thousands of leaves for caterpillars to feed on.

9 Draw a pyramid of numbers for each of the following food chains:

a 100 lettuces → 10 000 slugs → 100 thrushes → 1 hawk

b 1 rose bush → 10 000 greenfly → 1000 ladybirds

You should now have an understanding of these key ideas. You should also be able to spell and know the meaning of the key words. The key words are in **bold** on this page.

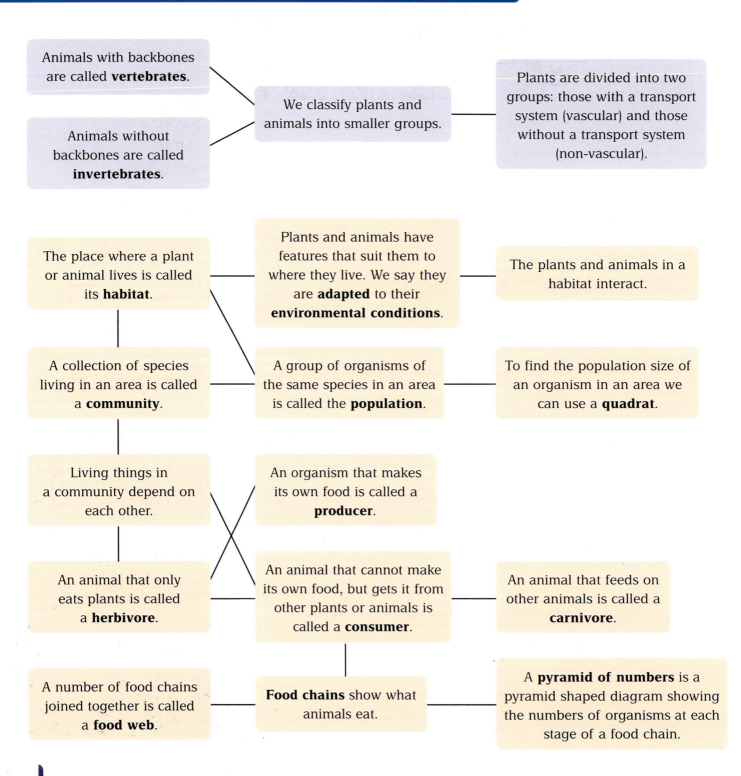

Animals with backbones are called **vertebrates**.

Animals without backbones are called **invertebrates**.

We classify plants and animals into smaller groups.

Plants are divided into two groups: those with a transport system (vascular) and those without a transport system (non-vascular).

The place where a plant or animal lives is called its **habitat**.

Plants and animals have features that suit them to where they live. We say they are **adapted** to their **environmental conditions**.

The plants and animals in a habitat interact.

A collection of species living in an area is called a **community**.

A group of organisms of the same species in an area is called the **population**.

To find the population size of an organism in an area we can use a **quadrat**.

Living things in a community depend on each other.

An organism that makes its own food is called a **producer**.

An animal that only eats plants is called a **herbivore**.

An animal that cannot make its own food, but gets it from other plants or animals is called a **consumer**.

An animal that feeds on other animals is called a **carnivore**.

A number of food chains joined together is called a **food web**.

Food chains show what animals eat.

A **pyramid of numbers** is a pyramid shaped diagram showing the numbers of organisms at each stage of a food chain.

Atoms and elements

In this unit we shall be learning about atoms and elements, and how atoms can join together to make compounds.

KEY WORDS
material
element
atom
particle
symbol
molecule
compound
formula

8E.1 Materials

Early ideas about elements

The photograph shows a **material**. You need to remember that a <u>material</u> is not the same as an <u>object</u>. Think about an ordinary ruler. The ruler is an object, and it is made out of plastic, which is a material.

We group materials in many different ways:

- natural and made;

- metals and non-metals;

- ceramics;

- fibres and plastics.

The Ancient Greeks thought that all materials were made up of four basic materials – Earth, Fire, Air and Water. They called them the four 'Elements'. For example, liquids such as wine or oil were thought to be mostly 'Water', and solids such as bread or iron were thought to be mostly 'Earth'.

Now we know that Earth, Fire, Air and Water are not elements at all, and that there are many more than just four elements. Our idea of what an element is has changed completely.

Wood is the <u>material</u> which makes up this pepper mill. The pepper mill is an <u>object</u>. Wood is a natural material.

1   What were the four 'Elements', according to the Ancient Greeks?

Elements today

Today we know that there are about 100 **elements**, and that there is something very special about them. Every element is a pure substance, and no element can be made out of anything else.

When you look at some materials with a microscope you can <u>see</u> that they are made up of different things. For example, concrete contains pieces of sand and different coloured crystals. Materials which contain several different things cannot be elements, so it is easy to see that concrete is not an element. Sometimes it is not so easy to tell. For example, glass, rubber and polythene are not elements. You just can't see that they contain different things.

Diamond is made of carbon only. Carbon and gold are elements.

Concrete is not an element. It is made of different things.

2   What is so special about elements?

3   How can you be certain that wood is not an element?

Arab, and then European scientists, called alchemists continued where the Greeks left off. They studied and tried to change materials. By the 18th century, scientists realised that some of the new materials they discovered, such as phosphorus, hydrogen and oxygen, were themselves elements.

These scientists recognised that every element was a pure substance. They were thinking of elements in the same way as we think of them today. One aim of many of the alchemists was to turn lead into gold. Eighteenth-century scientists started to understand that no element could be made out of anything else.

Wood is not an element. It contains cells of different shapes, sizes and colours.

Remember: every element is a pure substance, and no element can be made out of anything else.

Now we know that all the countless different materials in the world are made from the 100 or so elements. We think of elements as the 'building blocks' of nature from which everything is made, including us.

4   There are just over 100 different elements. Find out the names of <u>10</u> elements you have not heard of before.

8E.2 What the elements are like

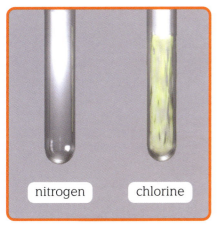

nitrogen   chlorine

gold

carbon

silicon

iodine

mercury

sodium

sulphur

All these are elements.

There are about 100 different elements. Some elements, such as gold and copper, have been known since prehistoric times. Other elements, such as radium, have only been known for about 100 years. Scientists are quite sure now that they have discovered all the <u>natural</u> elements. Scientists think it may be possible to make more new elements artificially.

Unlike materials such as wood, oil and soil, elements cannot be made up from anything else. This is because elements contain just one thing – themselves. All the other materials in the world are made from these 100 or so elements.

1   About how many elements are there?

Introducing atoms

Over 2000 years ago a Greek philosopher called Democritus suggested that everything was made of uncuttably small **particles**, which he called '**atoms**' (in Greek 'atomos' means 'uncuttable'). Many centuries later the idea became popular again, and in the nineteenth century scientists such as John Dalton began to find evidence for the existence of atoms. Because atoms are so small – too small to see even with a microscope – it was not easy to convince everybody that they existed.

Now scientists think that all materials are made up of particles that are too small to see. You learned in Unit 7G that it is the way the particles in a material are arranged that makes a particular material a solid, a liquid or a gas.

solid

liquid

gas

The arrangement of particles in solids, liquids and gases.

2   How are the particles arranged in a gas?

3   What does the word 'atom' mean?

Evidence for the existence of particles is all around us, and without the idea of particles there would be an endless number of things we could not explain. If you polish a piece of aluminium with a cloth you can smell the metal. This is because particles of aluminium (aluminium atoms) have gone into the air – and up your nose!

Atoms are one type of particle. We now know that every element has its own special kind of atom. There are oxygen atoms, iron atoms, carbon atoms, aluminium atoms, and so on. Each kind of atom is different from other kinds in its size and its properties.

You can smell the aluminium – this is evidence for the existence of aluminium atoms.

4   How does your nose provide evidence that particles exist?

We usually picture atoms as being tiny spheres, like marbles or ball-bearings. Imagine a large number of Lego pieces, all of the same colour, stuck together to make a huge Lego lump. It's a bit like this with atoms – you can arrange them to make different shapes. This is how we can get crystals of different shapes.

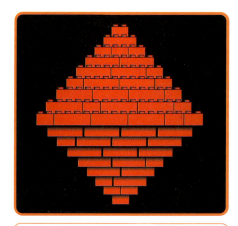

This large 'lump' of Lego is made from lots of identical small Lego pieces stuck together.

These iodine crystals are made from millions of iodine atoms held together.

Sodium chloride crystals are made up of sodium particles and chloride particles held together.

5 What type of atoms would you find in a lump of the element iron?

We now have a good idea of the size of atoms and molecules. Hydrogen atoms, the smallest of all, are about 0.000 0001 mm across.

6 About how big are the atoms that make up hydrogen?

7 Do you think carbon dioxide is an element? Explain your answer.

Tables of elements

The table shows some of the common elements. It shows the chemical symbol for each element and some other useful information.

Name	Symbol	Metal or non-metal	Solid, liquid or gas at 20°C	Colour	Year discovered
bromine	Br	non-metal	liquid	brown	1826
calcium	Ca	metal	solid	grey	1808
carbon	C	non-metal	solid	black	ancient
chlorine	Cl	non-metal	gas	green	1810
cobalt	Co	metal	solid	grey	1739
copper	Cu	metal	solid	pink	ancient
gold	Au	metal	solid	gold	ancient
helium	He	non-metal	gas	colourless	1868
hydrogen	H	non-metal	gas	colourless	1783
iron	Fe	metal	solid	grey	ancient
magnesium	Mg	metal	solid	grey	1808
mercury	Hg	metal	liquid	silver	ancient
nitrogen	N	non-metal	gas	colourless	1772
oxygen	O	non-metal	gas	colourless	1774
phosphorus	P	non-metal	solid	white	1669
silver	Ag	metal	solid	silver	ancient
sulphur	S	non-metal	solid	yellow	ancient

People have known about some elements, such as copper, silver and gold, for a long time. These elements already had common, everyday names in most languages. As scientists discovered new elements, they had to give them names. They chose the names for the new elements with great care and often tried to make their names fit what they were like. For example, the name 'chlorine' comes from a Greek word 'chloros' meaning green, and 'oxygen' means 'acid-maker'.

Every element has a **symbol** as well as a name. Some symbols are single letters, like O for oxygen; others have two letters, as in Au for gold. When there are two letters, the first is always a capital letter and the second is always a small letter. This is very important: Cu means copper but CU means carbon with uranium – something very different!

> The Periodic Table contains information about all the elements. You may have one in your science laboratory.

Scientists use the symbol to represent one atom of an element, and use a small number below the line if there is more than one atom. So, Fe means 'one atom of iron' and O_2 means 'two atoms of oxygen'.

8 What does Cl_2 mean?

9 What is the difference between CO and Co?

10 Which of the following are <u>not</u> atoms?

   a O_2 **b** H_2 **c** He

   d CO_2 **e** C

11 Find out about the discovery of helium and where its name comes from.

Introducing molecules

Atoms are particles and so are **molecules**. Molecules are groups of atoms stuck together in twos, threes, and so on; some molecules contain hundreds of atoms.

Sometimes a molecule contains two atoms that are the same. An example is an oxygen molecule, which is shown as O_2. However, there may be many different kinds of atom combined in a molecule. For example, carbon dioxide contains two different kinds of atoms and ammonium chloride contains three different kinds of atoms (the 'ammonium' bit contains hydrogen and nitrogen).

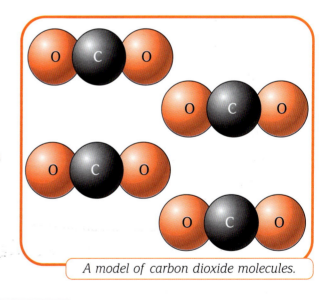

A model of carbon dioxide molecules.

12 What is the minimum number of atoms in a molecule?

8E.3 How we get all the other materials

There are two types of changes: <u>physical changes</u> and <u>chemical changes</u>.

Physical changes are things like melting and freezing, expanding and contracting. It is usually quite easy to reverse a physical change.

Just put the water back in the freezer to reverse this physical change.

Chemical changes are very different. The important difference is that it is usually very difficult, and sometimes impossible, to reverse a chemical change. This is because chemical changes actually make <u>new</u> materials, and do not just change the same material into a different form.

Burning is an example of an <u>irreversible</u> change you will already have met. Burning is only one example; here are some more:

- digesting food;
- leaves rotting away;
- iron rusting;
- cooking food;
- concrete setting.

Burning is an irreversible chemical change.

1 Give <u>one</u> example of a physical change and <u>one</u> example of a chemical change.

2 Looking at the remains of a bonfire gives a strong clue that burning is a chemical change. What is this clue?

Making new materials

We use chemical changes to get thousands of different materials from a limited number of starting materials. Sometimes materials can be put together to make new materials:

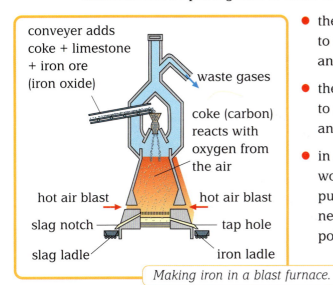

conveyer adds coke + limestone + iron ore (iron oxide)

waste gases

coke (carbon) reacts with oxygen from the air

hot air blast   hot air blast

slag notch   tap hole

slag ladle   iron ladle

Making iron in a blast furnace.

- the Romans knew how to make soap from soda and fat;

- the Romans knew how to make glass from soda and sand;

- in a modern chemical works, oil products are put together to make new materials such as polystyrene.

Making soap, the Roman way.

Polystyrene is made from oil.

When elements are combined together they form **compounds**. In most cases, what a compound is like does not seem to give any clue about the elements it contains. For example:

- sodium is a very reactive silvery metal;

- chlorine is a poisonous green gas.

Yet, sodium combines with chlorine to form sodium chloride, which is a harmless white solid known as common salt. Remember that we can think of a piece of an element as a huge Lego lump made up of millions of exactly the same type of brick. In the same way, we can think of compounds as being made up from two, or three, or more different types of brick. There will be a basic pattern, repeated again and again.

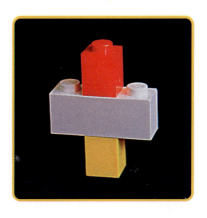

This model could represent chalk. The basic pattern, 1 red + 3 white + 1 yellow, represents the basic structure of the compound.

3 Water is a compound made from the elements hydrogen and oxygen. How is water different from the elements it contains?

4 What is the smallest number of different kinds of element a compound can contain?

5 Find out how sodium and chloride particles are arranged in a crystal of sodium chloride. Include a diagram in your answer.

8E.4 Representing the changes

Each compound is known by its **formula** and its name. For example, <u>carbon dioxide</u> is a compound. Each molecule of carbon dioxide contains one atom of carbon and two atoms of oxygen. We can write the formula of carbon dioxide as CO_2.

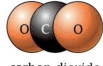

carbon dioxide

> The plural of 'formula' is 'formulae'.

Chemical names of compounds are a description of what is in them. For example, copper oxide contains copper and oxygen. We use the symbols of these elements to write the formula, so the formula of copper oxide is CuO.

Sometimes, as with carbon dioxide, the formula has small numbers below the line to tell you the ratio of each element present. CO_2 means that carbon dioxide has one atom of carbon for every two atoms of oxygen.

Every compound has a 'proper' chemical name, even if it already has an 'everyday' name. For example, common salt is sodium chloride and chalk is calcium carbonate.

The ending to the name is important. Names that end in 'ide' have two elements present. Names that end in 'ate' have three elements present including oxygen. So, calcium carbonate is made up of calcium, carbon and oxygen.

'Pass the sodium chloride please.'

Calcium carbonate.

1 What is the difference between the everyday name for a compound and its chemical name?

2 What elements do you think these compounds contain?

  a copper chloride

  b zinc sulphide

  c nitrogen oxide

3 These are the formulae of real compounds. Work out the name for each of them.

  a CaO

  b MgS

  c HCl

More formulae and equations

Compounds may have different names in different languages, but their formulae are always the same. So a scientist, in any country, can recognise a compound by its formula.

Scientists can work out the formula for water by splitting water into hydrogen and oxygen. This shows that water contains twice as much hydrogen as oxygen. So, the formula of water is H_2O.

We represent chemical reactions with <u>chemical equations</u>. We write them using the names (a word equation), using formulae (a symbol equation), or both!

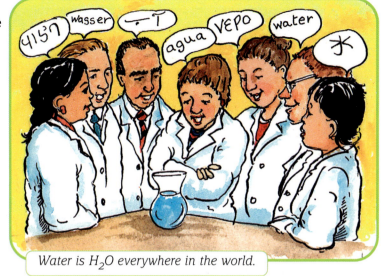

Water is H_2O everywhere in the world.

$$C \quad + \quad O_2 \quad \rightarrow \quad CO_2$$

carbon   oxygen   carbon dioxide

We can also add information about the chemicals we start with, the <u>reactants</u>, and the chemicals we make, the <u>products</u>.

$$2Mg \quad + \quad O_2 \quad \rightarrow \quad 2MgO$$

magnesium   oxygen   magnesium oxide

<u>reactants</u>   <u>product</u>

In the two equations above, O_2 represents a <u>molecule</u> of oxygen gas. An oxygen molecule is made up of two oxygen atoms joined together. This sort of molecule is called a diatomic molecule. Many of the common gases, including hydrogen, nitrogen and chlorine, have molecules of this kind.

4   What is a diatomic molecule?

5   Write a word equation to show copper reacting with sulphur to make copper sulphide.

6   Calcium reacts with oxygen. Write a word equation for this.

7   Look at the equation for the reaction between magnesium and oxygen. Explain why we need two atoms of Mg and why we make two lots of MgO as the product.

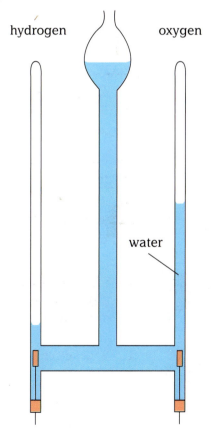

One way of splitting up water is by using electricity – this process is called electrolysis. A Hofmann voltameter is used for the electrolysis of water. Twice as much oxygen is produced as hydrogen.

You should now understand the key words and key ideas shown below.

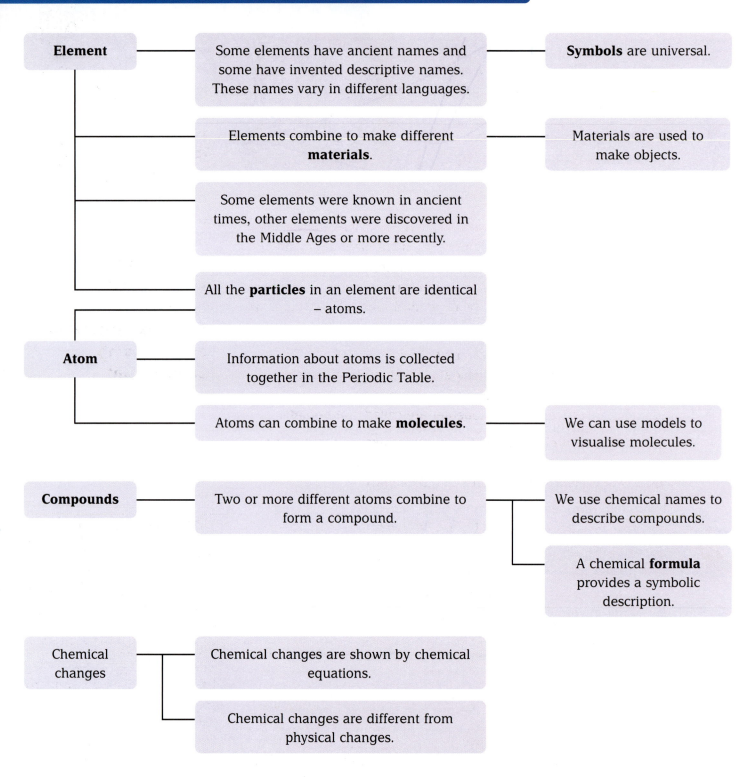

Element — Some elements have ancient names and some have invented descriptive names. These names vary in different languages. — **Symbols** are universal.

Elements combine to make different **materials**. — Materials are used to make objects.

Some elements were known in ancient times, other elements were discovered in the Middle Ages or more recently.

All the **particles** in an element are identical – atoms.

Atom — Information about atoms is collected together in the Periodic Table.

Atoms can combine to make **molecules**. — We can use models to visualise molecules.

Compounds — Two or more different atoms combine to form a compound. — We use chemical names to describe compounds.

A chemical **formula** provides a symbolic description.

Chemical changes — Chemical changes are shown by chemical equations.

Chemical changes are different from physical changes.

Compounds and mixtures

In this unit we shall be looking at elements and compounds in more detail. We shall also look at mixtures and see how they differ from both pure elements and compounds. In particular, we shall look at an important mixture that is all around us – the air!

KEY WORDS

atoms
elements
compounds
react
mixture
oxygen
nitrogen
argon
carbon dioxide
water vapour
liquefied
fractional distillation
melting point
boiling point

8F.1 Elements and compounds

In Unit 8E, you learnt that all substances are made up of small particles called **atoms**. Some pure substances are made up of only one type of atom. These pure substances cannot be broken down into anything simpler by a chemical reaction. We call these substances **elements**.

There are only 92 different elements found naturally on Earth. Atoms of these elements join together and make millions of different substances. Substances formed from two or more elements chemically joined together are called **compounds**.

We can think of elements and compounds like we think of letters and words. There are only 26 letters in the alphabet but we can use them to make all the words in a dictionary.

1 Read the passage about noble gases, then:

 a count the number of different words used;

 b find out how many different letters are used;

 c write down the words that contain only one letter;

 d find the word that has the greatest number of different letters.

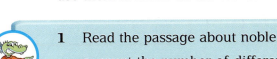

2 Why did the inert gases go undetected for centuries?

Many elements are well-known substances. Oxygen, iron and copper are examples. However, some elements are very rare and you do not see them very often. Samarium is an example.

There are inert gases in the air we breathe. They have strange Greek names which mean "the New", "the Hidden", "the Inactive" and "the Alien". They are so satisfied with their condition that they do not join with any other element. Only forty years ago did a chemist finally succeed in forcing the Alien (xenon) to combine briefly with the element fluorine.

Now, we often call inert gases the noble gases.

3 Look at the pictures of one use of each of four elements. Which element:

   a is a shiny metal used to build bridges?

   b is a gas in the air, that you can't do without?

   c is a gas used in airships?

helium   oxygen   iron

We cannot see atoms because they are very small. However, we can represent what an element or compound looks like by using models. Models will show which atoms are found in the substance.

4 In the particle diagram, which substances are elements and which substances are compounds?

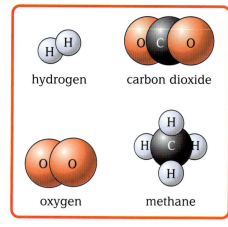

hydrogen   carbon dioxide

oxygen   methane

We can describe each of these substances using a chemical formula. A chemical formula tells you how many atoms of each element there are in the smallest particle of the substance.

5 How many atoms does one particle of water contain?

6 Work out the formula for each substance shown in the particle diagram.

7 Look at the formulae given on the bottles. For each formula work out:

   a how many atoms it contains;

   b how many different elements are present.

8 The labels for these compounds have fallen off the bottles. Match the names of the compounds to the correct bottles.

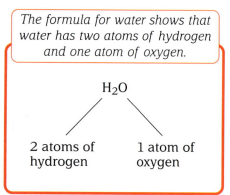

The formula for water shows that water has two atoms of hydrogen and one atom of oxygen.

$$H_2O$$

2 atoms of hydrogen   1 atom of oxygen

H_2SO_4 $CaCO_3$ MgO $CuSO_4$

calcium carbonate   sulphuric acid

copper sulphate   magnesium oxide

8F.2 Looking at compounds

<u>Sodium chloride</u> is a compound made from the elements <u>sodium</u> and <u>chlorine</u>.

When these two elements meet, they **react** violently to produce white crystals of sodium chloride. It is a new substance made from two different elements, so it is a compound. We know that the reaction is a chemical reaction because salt is very different from both sodium and chlorine. Compounds are always different from the elements they are made from.

Sodium is a reactive metal that will react violently with water to produce hydrogen.

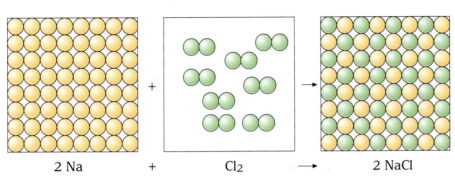

2 Na   + Cl_2 ⟶   2 NaCl

This particle diagram shows the reaction between sodium and chlorine.

Chlorine is a pale green gas which is very poisonous.

1   What is the chemical symbol for sodium?

2   What is the chemical symbol for chlorine?

3   What is the chemical formula for sodium chloride?

We can describe what happens in this chemical reaction by writing a word equation. In a word equation the reactants are shown on the left-hand side of the arrow and the products are shown on the right-hand side of the arrow.

sodium + chlorine → sodium chloride
    reactants   product

Sodium chloride is the salt that we add to our food to give it a salty taste.

4   Look at the diagrams. Describe the appearance of iron and of sulphur.

The word equation for the reaction between iron and sulphur is:
iron + sulphur → iron sulphide

5   Describe the differences between iron sulphide and:

   a iron;

   b sulphur.

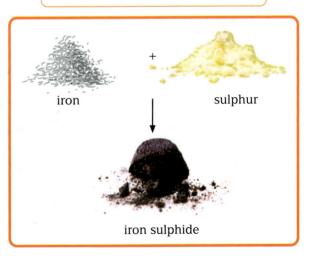

iron   sulphur

iron sulphide

8F.3 Compounds and some of their reactions

Some compounds are very reactive. However, some compounds are less reactive and will only react under certain conditions.

1 Look at the pictures. Describe the evidence they give for chemical reactions.

In earlier units you saw many different types of reactions, including:

- reactions with oxygen. Words used to describe reactions with oxygen include 'burning', 'combustion', 'oxidation' and even respiration;

- neutralisation reactions. When an acid reacts with a base, the properties of the acid and base are cancelled out;

- precipitation reactions. In some reactions between two solutions, one of the products is insoluble and it settles as a solid called a <u>precipitate</u>;

- thermal decomposition reactions. Some compounds break down or <u>decompose</u> when you heat them. For example, calcium carbonate is broken down into the simpler substances calcium oxide and carbon dioxide when it is heated.

2 Oxygen reacts with many different substances. Where does the oxygen come from for all these reactions?

3 Find out how you can tell when:

a an acid has been neutralised;

b a gas is produced in a neutralisation reaction.

4

| sodium carbonate solution | + | iron chloride solution | → | iron carbonate precipitate | + | sodium chloride solution |

For this reaction, explain how to separate the products.

5 Write down the word equation for the thermal decomposition of calcium carbonate.

You know from Unit 7F that a chemical reaction may have taken place if:

- a gas is produced;

- a colour change happens;

- heat is produced;

- there is a change in mass;

- there is a change in appearance.

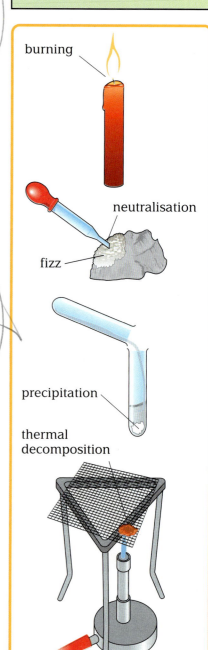

burning

neutralisation

fizz

precipitation

thermal decomposition

8F.4 What is a mixture?

A pure element is made up of identical particles, all of which are made up of the same type of atom.

A pure compound is made up of identical particles, but each particle contains two or more different types of atom chemically joined together.

However, many substances are mixtures. A **mixture** contains more than one substance and therefore it will contain more than one type of particle.

Look at the particle diagrams. Many substances are mixtures. There are different types of mixtures. You can have mixtures of:

● elements;

● compounds;

● elements and compounds.

1   Look at the particle diagram of a mixture. How many different types of particles are there in this mixture?

2   Which of the particles present in the mixture are:

   a elements;

   b compounds?

3   Choosing your own elements and compounds, draw particle diagrams to show mixtures of:

   a two elements;

   b an element and a compound;

   c two compounds.

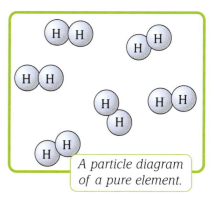

A particle diagram of a pure element.

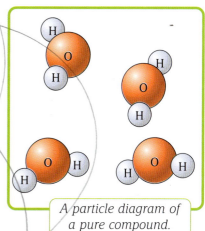

A particle diagram of a pure compound.

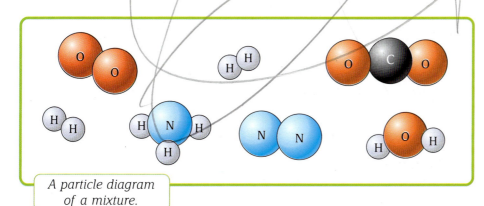

A particle diagram of a mixture.

You will come across many examples of mixtures. For example, mineral water is a mixture. Look at the label.

4 How many different minerals are there in the mineral water?

5 Which mineral is:

a the most common?

b the least common?

6 Why can't you see the minerals in the water?

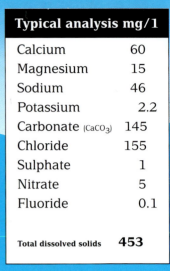

Typical analysis mg/1	
Calcium	60
Magnesium	15
Sodium	46
Potassium	2.2
Carbonate (CaCO$_3$)	145
Chloride	155
Sulphate	1
Nitrate	5
Fluoride	0.1
Total dissolved solids	**453**

Sea water is also a mixture. It contains many different solids, called salts, dissolved in the water. There is an average of 40 g of salts dissolved in every kilogram of water from an ocean.

But there are about 370 g of salts dissolved in every kilogram of water from the Dead Sea. The Dead Sea is the world's saltiest natural lake. It is an important mixture and is one of the greatest sources of minerals in the world.

Over 43 billion tonnes of salts are thought to be dissolved in the Dead Sea, of which almost 2 billion tonnes are potassium chloride.

The water in the Dead Sea contains so many dissolved salts that you can easily float in it.

Salts in the Dead Sea

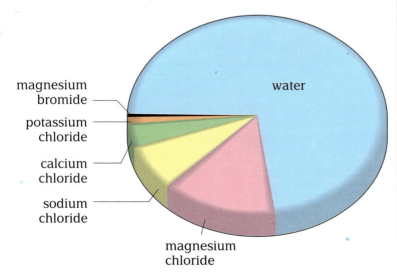

Mineral	% mass
magnesium chloride	14.5
sodium chloride	7.5
calcium chloride	3.8
potassium chloride	1.2
magnesium bromide	0.5
water	72.5

Mineral	Amount (billion tonnes)
magnesium chloride	22
sodium chloride	12
calcium chloride	6
potassium chloride	2
magnesium bromide	1

7 What is the most common compound in the Dead Sea mixture, after water?

8 What is the mass of calcium chloride in the Dead Sea?

The water going into the Dead Sea is not particularly salty. However, the local climate is extremely hot and dry, so the rate of evaporation of water is high. Water evaporates and leaves the salts behind. The Dead Sea is a bit like a huge evaporating basin.

9 **a** How many grams of salts are dissolved in every kilogram of water in the Dead Sea?

b How many times greater is this than the amount of salts dissolved in an ocean?

c How do all these salts get into the Dead Sea?

10 Why is the Dead Sea important for industry?

Air – a very special mixture

The air we breathe is a mixture of a number of very important gases.

Gas	Percentage in air
oxygen	21
nitrogen	78
argon (and other noble gases)	1
carbon dioxide	0.035
water	6 – 0.1

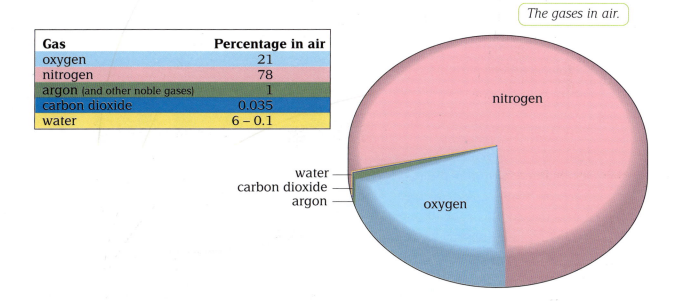

The gases in air.

nitrogen

oxygen

water
carbon dioxide
argon

11 Which gas makes up the biggest part of the mixture we call air?

12 What is the percentage of carbon dioxide in the air?

The most important gas in the air is the one that we all need to keep us alive – **oxygen**. We use oxygen to release the energy we need to keep our bodies working. We get our oxygen from the air and it is absorbed into our bloodstream through our lungs.

In hospital, patients with breathing or heart problems are sometimes given pure oxygen to breathe, rather than air. This makes it easier for them to absorb oxygen into their blood.

Oxygen is used in industry to produce important substances such as steel and nitric acid. Welders use oxygen to produce a very hot flame.

Nitrogen is an unreactive gas. This means we can use it to cool or store substances that react with oxygen, such as frozen food, chemicals and electronic equipment. We also use nitrogen to make the nitrate fertilisers that help crops grow.

There is a mixture of noble gases in the air but **argon** is the most common. Noble gases are very unreactive so they have little use in making new substances. Argon is the gas used inside light bulbs because it does not react with the tungsten filament in the bulbs.

Noble gases are used to produce bright lighting in advertising signs.

13 Write down:

   a <u>two</u> uses of oxygen;

   b <u>two</u> uses of nitrogen;

   c <u>one</u> use of noble gases.

We use **carbon dioxide** in some kinds of fire extinguishers, as it is very good at stopping things burning. Carbon dioxide is also the gas found in fizzy drinks. Also, we use solid carbon dioxide to store frozen food.

We call solid carbon dioxide 'dry ice'. Dry ice turns straight from a solid to a gas.

The amount of **water vapour** in the air varies from day to day, depending on the weather. In some places the water vapour content can be as low as 0.1% but in a warm, humid climate it can be as high as 6%.

14 Draw a box filled with air showing all the different kinds of particles present.

Separating the gases in air

Air is a mixture of gases. Each gas in the mixture is a pure substance, so it has its own boiling point. In order to separate the gases present in air we must cool the mixture down. The first thing that happens is that water vapour condenses, and then solidifies. Then, as the air gets colder, carbon dioxide also becomes a solid.

15 At what temperature does water become a solid?

16 Solid carbon dioxide and solid water would cause problems when the air mixture passed through pipes in the equipment used to separate air. Why is this?

The air is now dry and doesn't have any carbon dioxide in it. Next, the air is squashed to a high pressure, then allowed to expand quickly. This cools the air very quickly and it turns to a liquid. We say that the air has been **liquefied**.

Then the liquid air is allowed to warm up. It starts to boil. As the air warms up, the substance with the lowest boiling point boils first and turns into a gas. This gas is collected and stored. As the temperature rises slightly, the substance with the next lowest boiling point boils, and is collected and stored separately. So the gases are separated.

Over 95% of the oxygen used in industry and medicine is obtained from liquefied air.

This process of separating a mixture of liquids by heating them and collecting the gases as they boil off is called **fractional distillation**. We use a similar process to separate crude oil into simpler substances such as petrol and diesel. We also use distillation to separate the alcohol from water when we make drinks like whisky.

Separating the gases in air.

AIR

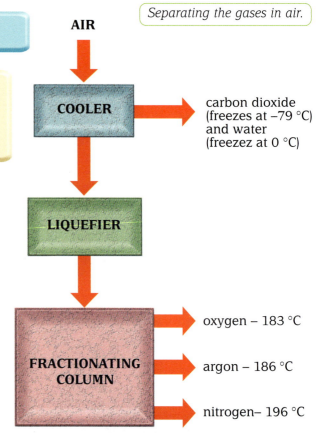

COOLER → carbon dioxide (freezes at −79 °C) and water (freezez at 0 °C)

LIQUEFIER

FRACTIONATING COLUMN → oxygen − 183 °C

argon − 186 °C

nitrogen− 196 °C

Liquid nitrogen.

17 Which substance in air has the highest boiling point?

18 Which substance in air has the lowest boiling point?

19 When liquefied air heats up, which is the first substance to become a gas?

20 Which substance in the liquefied air is the last to change back to a gas?

21 What is fractional distillation?

22 Write down <u>two</u> other uses of fractional distillation.

Boiling points of gases in air.

Gas	Boiling point
nitrogen	−196 °C
oxygen	−183 °C
argon	−186 °C
water	100 °C

Mixtures and pure substances

- Iron is an element.
- Water is a compound.
- Hydrogen is an element.
- Methane is a compound.

But they are all pure substances.

 23 At room temperature, which of these substances are:

    a solids   **b** liquids   **c** gases

 24 How can you turn methane into a liquid?

If you cool water down, it eventually becomes a solid when the temperature is 0 °C. We call this solid <u>ice</u>. All pure substances, whether they are elements or compounds, have a fixed **melting point**.

If you heat water up it will eventually start to boil. This happens at a fixed boiling point of 100 °C. All elements and compounds have a fixed **boiling point.**

 25 What is the fixed melting point of ice?
      26 What is the fixed boiling point of water?

We can use the melting point or boiling point to tell us if the substance is pure. If we add salt to water the mixture doesn't boil at 100 °C, but at about 106 °C. Also, the mixture doesn't melt at 0 °C, but at about –6 °C. It depends on how much salt we add.

 27 We put salt on roads in winter when a frost is forecast. Why does salt help keep the roads clear of ice?

Mixtures are not pure substances, so they do not have fixed melting points or boiling points. A mixture will melt or boil over a range of temperatures. The exact range will depend on what is in the mixture. We add antifreeze to the water in a car's cooling system to make sure that it does not freeze if the temperature goes below 0 °C.

 28 Why is it bad for the engine when its cooling water freezes?

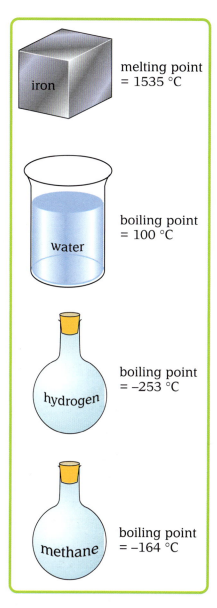

iron — melting point = 1535 °C

water — boiling point = 100 °C

hydrogen — boiling point = –253 °C

methane — boiling point = –164 °C

You should now understand the key words and key ideas shown below.

- An **element** contains only one type of **atom**.

- A **compound** contains different types of atoms joined together.

- The formula of a substance tells us the number of each type of atom present in the substance.

- The properties of a compound are different from those of the elements which it contains.

- A compound is a pure substance which is made of the same particles.

- A sample of a compound will always have the same elements present in the same fixed proportions.

- Compounds will **react** chemically to make new substances.

- You can tell that a compound has reacted by observing what happens to it.

- A **mixture** is formed when two or more pure substances are added together.

- Air is a mixture of gases that can be separated into pure substances.

- Air consists of **nitrogen**, **oxygen**, **argon** (and other noble gases), **carbon dioxide** and **water vapour**.

- Each of the gases found in air has important uses.

- Air can be **liquefied** and then the individual gases can be separated by **fractional distillation**.

- Sea water and mineral water are other examples of mixtures.

- Elements and compounds melt and boil at a certain temperature.

- Mixtures do not melt or boil at one particular temperature.

- The **melting point** and **boiling point** of a mixture change as the composition of the mixture changes.

Rocks and weathering

In this unit you will learn about rocks and how the materials that rocks are made of are recycled by weathering, erosion and deposition. These natural processes happen over a long period of time.

KEY WORDS
minerals
textures
porous
non-porous
weathering
limestone
chemical weathering
sedimentary rocks
sandstone
igneous rock
granite
erosion
deposition
sediments
abrasion
fossils
millions of years

8G.1 Rocks vary

Scientists who study rocks are called <u>geologists</u>. We sometimes say that something is 'as hard as a rock', but not all rocks are hard. Geologists call <u>all</u> the materials in the photographs rocks.

1 Sort the rocks into two groups; solid rocks and loose-grained rocks (soft like sand).

Most rocks are made of a mixture of grains of different sizes. The grains can be made of different substances which we call **minerals**.

2 Write down:

a <u>one</u> rock with a mixture of small and large grains;

b <u>one</u> rock with small grains;

c <u>one</u> rock that is clearly made of a mixture of minerals.

Geologists sort rocks into two groups, according to the way the grains fit together. They say that the two groups have different **textures**. In one group, the grains have small spaces between them, called pores, so we call them **porous** rocks. In the other group, the grains fit closely together with no spaces or pores. So we say that they are **non-porous**.

Look at the photographs.

3   Is granite or sandstone the porous rock? Explain your answers.

4   What do you think is in the pores of porous rocks?

5   In some places we can drill boreholes in rocks to get water or oil. Do we find water and oil in porous rocks or in non-porous rocks? Explain your answer.

6   Gas and oil rise through pore spaces in rocks. Geologists searching for gas and oil look for structures that act as 'traps'. These traps stop the gas and oil moving up further. Find out what these traps are like.

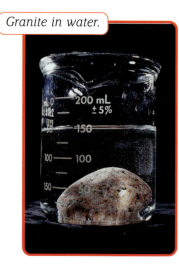

Granite in water.

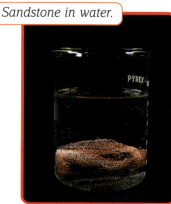

Sandstone in water.

8G.2 Rock and rain

Rocks don't stay the same for ever. Over a long period of time, rain, frost, temperature changes and wind all help to change rocks and break them down. Look at the photograph. The weather caused the change to the old gargoyle. So we call the process **weathering**.

1   a Describe the differences between the old and the new gargoyles.
   b How long did these changes take?

this gargoyle has just been replaced

this gargoyle is 500 years old

Explaining weathering

Rainwater is one of the main causes of weathering. You probably know that rain is slightly acidic. In Unit 7F, you learned that acids react with carbonates and with other substances.

- The gargoyles are made of **limestone**.
- Limestone is made of calcium carbonate.
- Acidic rain reacts with the calcium carbonate.
- One of the products is carbon dioxide. It escapes into the air.
- The other product dissolves, so it is washed away.
- So the limestone weathers.

The reaction between rain and the limestone is a chemical reaction. So we call this kind of weathering **chemical weathering**.

2 Explain where the missing bits of the gargoyle went.

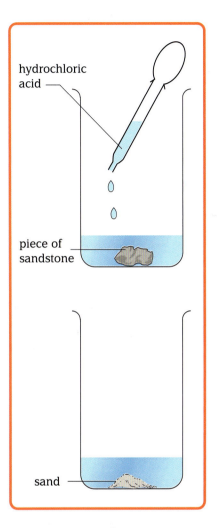
hydrochloric acid

piece of sandstone

sand

Limestones and sandstones are formed as loose deposits or sediments in the sea. They are called **sedimentary rocks**. Material called cement holds the grains together and makes the rock solid. In some types of **sandstones**, the grains of sand are held together with a cement of calcium carbonate. The pictures show what happens when you soak this kind of sandstone in acid.

3 Explain how chemical weathering turns some sandstones into a pile of sand.

4 The grains in the sandstone in the picture fell apart in about half an hour. Explain why it takes rain many years to break down sandstone.

Weathering very hard rocks

Deep down in the Earth, it is very hot. Sometimes the rocks there get so hot that they melt. We call the melted rocks magma. When magma cools, it becomes solid rock again. We call this new kind of rock an **igneous rock**. **Granite** is an igneous rock.

5 Look closely at the minerals in granite.
- quartz: semi-transparent (glassy) crystals;
- feldspars: large pink or white crystals;
- mica: small, black crystals.

Draw and label <u>one</u> crystal of each mineral.

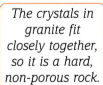

The crystals in granite fit closely together, so it is a hard, non-porous rock.

Rainwater reacts with feldspars and micas and changes them into new substances. This is chemical weathering. Some of the products are soluble so they dissolve in the water and are washed away.

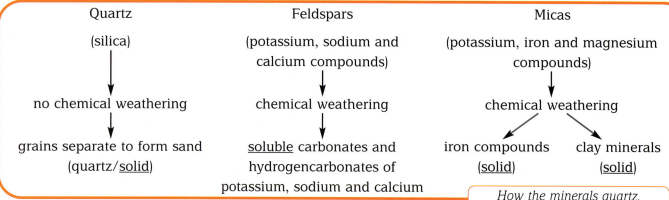

Quartz	Feldspars	Micas
(silica)	(potassium, sodium and calcium compounds)	(potassium, iron and magnesium compounds)
↓	↓	↓
no chemical weathering	chemical weathering	chemical weathering
↓	↓	↙ ↘
grains separate to form sand (quartz/<u>solid</u>)	<u>soluble</u> carbonates and hydrogencarbonates of potassium, sodium and calcium	iron compounds (<u>solid</u>) clay minerals (<u>solid</u>)

How the minerals quartz, feldspars and micas in granite are weathered by rain.

You don't need to learn the names of these chemicals, but you do need to know that rain weathers the three minerals in different ways.

6 Look at the chart for the weathering of granite. Write down:

   a <u>two</u> minerals that rain changes by chemical weathering;

   b <u>two</u> products of chemical weathering that are washed away.

7 **a** When granite weathers, which <u>two</u> solids are left behind?

   b Which one is in the same form as in the original granite?

8 Look at the photograph of the gargoyles on page 76. If these statues were made of granite, you would be able to feel little hollows where one of the minerals was washed away.

   Explain this as fully as you can.

9 Basalt is another hard, igneous rock. Look at the table.

A thin section of basalt

Mineral	Approximate % of mineral in	
	Granite	**Basalt**
quartz	38	3
feldspars	47	57
micas and other iron and magnesium minerals	15	40

Tanya soaked 100 g blocks of granite and basalt in concentrated acid for a month. Which one left the largest mass of solids at the end? Explain your answer.

8G.3 Rocks and temperature changes

What happens when water in rocks freezes?

When water changes to ice it takes up more space. We say that it <u>expands</u>.

The water in this bottle changed to ice.

1 Look at the picture. What else happened when the water turned to ice?

Most rocks have cracks in them that can fill up with water. When this water freezes, it expands. This means that the cracks widen. The change is very small, so it is hard to tell that it has happened. But when the ice melts, the rock breaks into pieces.

a *Look at the cracks in this shale.*

b *This limestone has pores and cracks.*

2 Explain how water freezing in cracks in rocks causes weathering.

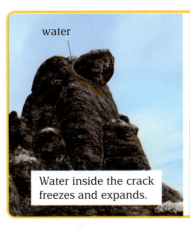

water

Water inside the crack freezes and expands.

As it gets warmer, the ice melts and bits of rock break off.

3 It is safer for mountaineers to climb early in the morning than at midday. Explain this as fully as you can.

4 Mountain peaks and the rocks that fall from mountains are not rounded. They have sharp angles. Why do you think this is?

5 Explain how the screes in the Alps formed.

Some screes in the Alps

What happens as rocks heat up and cool down?

Expansion of the rock itself causes weathering. If you have studied Unit 8I, you will know that:

- when we heat up a substance, the particles in it move faster and take up more space. So the space that the whole substance takes up increases – we say that it expands.

- when a substance cools down, the particles move more slowly and take up less space. The substance contracts.

6 The particles in a solid are close together.
Draw diagrams to show the difference between the particles in an expanded solid and a contracted solid.

In sunny places, rocks get hot during the day and cool down at night. So they expand and contract over and over again.
The large forces resulting from expansion and contraction make the rock crack.

7 Look at the picture. Which part of the rock gets hottest?

At the surface the rock
• heats up first;
• cools down first;
• expands the most.

Just below the surface the rock
• heats up less;
• cools more slowly.

In the middle of the rock, the temperature of the rock hardly changes.

8 a Copy the outline of the rock in the picture. Shade the part of the rock that you think will crack.

b Explain why you chose that part.

c Draw some cracks in red.

9 The kind of weathering shown in the picture is sometimes called <u>onion skin weathering</u>.
Why do you think this is?
Hint: Think about peeling an onion.

Very hot days and very cold nights caused this weathering.

8G.4 Moving weathered pieces of rock

As a result of weathering, rocks break down into bits that we call <u>rock fragments</u>. Sometimes forces move them away from where they formed. We call this **erosion**.

Gravity makes loose rock fragments fall or roll down slopes. Gravity makes water flow downhill too. Rainwater washes down smaller rock fragments. Then water currents in rivers carry them away.

1   Weathering and erosion are different processes. Explain the effect of the two processes on a rock.

2   One force that causes erosion is gravity. Explain <u>two</u> ways that gravity helps to move rock fragments.

When rock fragments settle, we call this process **deposition**. We call the deposits **sediments**. The pictures show the sediments deposited at different stages of a river's journey.

3   Look at the pictures. Describe the sediments in each picture.

4   Write down <u>two</u> ways that the fragments upstream are different from the fragments lower down.

Sediments are not all the same size

The slope of a river bed changes along its course. Look at the diagram.

HILLS

⟶ gentler slope
⟶ current flows more slowly
⟶ current has less energy to carry grains
⟶ smaller grains are carried further

SEA

5   Why does the water flow more quickly in the hills?

6   **a** In what way does the speed of the current affect the size of the fragments that it can move?

   b Explain why the speed of the current has this effect.

7   In the hills, the fragments of rock that fall into streams are a mixture of different sizes. But the ones on the bed of the stream are all large. Where do the smaller fragments go?

8   The river deposits sand in one place and mud in another. It seems to sort the sediment grains into similar sizes. Describe how it does this.

In the hills, streams flow quickly. Streams carry smaller rock fragments away and leave the large fragments behind.

Further downstream, we see beaches made of pebbles.

On flatter land, the river flows more slowly so it deposits sand.

The river deposits fine sand and mud as it gets nearer to the sea.

Why the shapes of rock fragments change

When weathered rock fragments form, they are angular. As the water moves them along, the fragments knock against each other. Corners get knocked off and they become more rounded. The water lifts up smaller grains and dashes them against other fragments causing wear. We call this wear **abrasion**. It is a bit like sand blasting. In a river, the sand blasting effect makes the rock fragments smoother and more rounded.

9 What is abrasion?

10 What is the connection between the shape of rock fragments and the distance they travel?

11 Find out how we use abrasion <u>either</u> to clean old buildings <u>or</u> to polish decorative stones.

8G.5 Why sediments form layers

In the last topic, you learned that rivers move sediments and that the sediments settle when the water doesn't have enough energy to carry them. This means that sometimes a river carries larger particles than at other times.

1 These pictures show the same river. Write down the differences between the two pictures. Include differences in speed of flow, and size and number of particles carried.

a The river carries only very small particles of sediment when it is flowing slowly.

b The river is in flood. When it flows quickly it carries a lot of sediment.

2 Look at the beaker of water and sand. Why does the sand settle when you stop stirring?

3 Look at what happens when the particles are a mixture of sizes. Describe what you see.

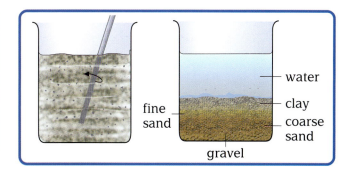

- water
- clay
- coarse sand
- fine sand
- gravel

4 Now look at the rocks in the cliff. Are the layers more, or less, clear than in the jar?

The distinct layers show that different sediments settle at different times.

How the layers in the cliff formed

The rocks in the cliff formed from sediments deposited in the sea. They are sedimentary rocks. When a river reaches the sea, the water slows down and it deposits its load of sediments.

- shale
- sandstone
- shale
- beach

Look at the picture. A layer of mud settled first. It formed the shale, so the shale is the oldest rock. It took many years to form. The top of this layer marks a time when deposition of sediment stopped. Later, deposits of sand formed the sandstone.

5 In the cliff, which is the youngest layer of rock and what is it made from? Explain your answers.

6 What do the layers of sand and mud tell you about the changes in speed of the river that brought them into the sea?

7 **a** Which of the layers of sediment probably took the longest to form? Explain your answer.

  b Why can't you be sure that you are right?

The rocks in the cliff are harder than new sediments in the sea. But the new sediments will harden in time. The weight of newer layers presses down on the older ones and squeezes water out. Dissolved solids from the water form crystals that cement the sediments together. The dissolved solids in rivers, lakes and seas come from weathered rock.

- sea
- mud
- sand
- mud

Sometimes rivers carry sand into the sea. At other times they carry mud.

8 Explain how the rock fragments that make up a sediment become cemented together.

Some rocks form mainly from dissolved solids

How dissolved solids become concentrated in seas and lakes.

rain

chemical weathering of rocks

dissolved salts in rivers

water evaporates leaving salts behind

sea

9 Look at the diagram.

a What kind of weathering produces dissolved solids?

b How do these substances reach the sea?

c Why is the concentration of dissolved solids much higher in the sea than in rivers?

Water evaporates from this soda lake faster than it flows in. As water evaporates, crystals of salts are deposited.

10 What makes the water evaporate from seas and lakes?

11 Salt deposits form in some seas and lakes but not in others.

Explain this as fully as you can.

Living things form rocks too

Many animals take dissolved solids out of the water to make their shell or skeleton. When the animals die, these hard parts don't decay; they form sediments too.

Corals and molluscs are two groups of animals that make their hard parts from calcium carbonate.

You can still see the shells and other hard parts of animals in these limestones. We call them **fossils**.

12 Write down <u>two</u> animals that concentrate dissolved solids.

13 Write down the name of the chemical that forms the hard parts of these animals.

14 a What happens to the soft parts of animals when they die?

b What happens to the hard parts when the animals die?

c What kind of sedimentary rock do they form?

Stories in rocks

When we know what to look for, rocks can tell us something of the history of a place over **millions of years**. But it wasn't until the 18th and 19th centuries that people started to realise this.

James Hutton, a British geologist who lived from 1726 to 1797, had a theory that rocks formed in the past in the same ways that they form today. We can use this idea to work out whether a rock formed in the sea, in a lake or in a desert. For example, corals live only where the sea is clear, warm and fairly shallow. So rocks that contain corals must have formed in the same conditions.

In a river delta mud is carried to the sea and settles in layers. This photograph was taken from a Space Shuttle.

15 If you find a fossil coral in a rock, what does this tell you about the environment at the time the rock formed?

16 How do we know that the rock in the photograph formed in a delta?

This rock formed in a river delta. The stripes in the rock are called <u>current bedding</u>.

Fossils also help to tell the story

Fossil collecting was popular in the nineteenth century. Mary Anning was one of the best-known fossil hunters. She, and her brother Joseph, found the first complete fossil of an ichthyosaur.

William Smith used fossils to help him to make the first geological map of England and Wales. His idea was that rocks with the same fossils in them are the same age. So he used fossils to match rocks in different parts of the country to help him to make his maps.

17 Another of William Smith's ideas was, 'Of any two strata [layers of rock], that which was originally below is the older.' What did he mean?

Telling the story

Look at the drawing. The rocks in the cliff took millions of years to form.

18 Draw a diagram to show the layers in the cliff face. Write on the diagram the names of the rocks. Then label:

  a the oldest layer;

  b a layer formed in a clear sea;

  c a layer formed when a river brought lots of sand into the sea;

  d a layer formed when the sea dried up;

  e a layer formed in a river delta.

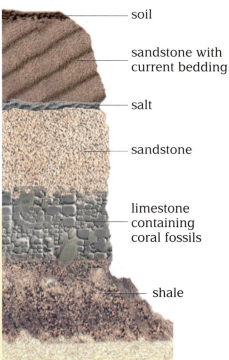

soil

sandstone with current bedding

salt

sandstone

limestone containing coral fossils

shale

You should now understand the key words and key ideas shown below.

ROCKS are made of a mixture of **mineral** grains.

We call the way grains fit together the **texture**.

- In a **non-porous** rock, grains fit closely together.
- In a **porous** rock, they have spaces between them.

Rocks such as **sandstone** and **limestone** form when sediments settle. They are **sedimentary** rocks.
Rocks such as **granite** form when magma solidifies.
They are **igneous** rocks.

Weathering breaks down rocks.

Water causes **chemical weathering** of some minerals.

Mechanical weathering is caused by:

- temperature changes,
- water expanding as it freezes.

We call the transport of rock fragments **erosion**.

The faster a river flows, the larger the fragments that it can carry.

The further rock fragments travel, the smaller and smoother they become. Wearing of rocks by sand is called **abrasion**.

When rock fragments settle and form layers, we call it deposition.

Sediments are **deposited** as rivers slow down.

When a river reaches the sea, it slows down so much that even the smallest rock fragments are deposited.

The boundary between two layers of rock represents a time when deposition stopped.

The remains of dead organisms (**fossils**) sometimes form deposits.

Evaporation of water causes deposition of dissolved solids.

Remember:

- When we study the Earth we can look at processes that are happening now, and work out what happened in the past.
- These processes are very slow. They take place over **millions of years**.

The rock cycle

In this unit you will be finding out more about different types of rocks, how rocks form and how the materials in rocks are used over and over again, both on the Earth's surface and deep under the ground.

KEY WORDS
sediment
sedimentary rocks
fossils
metamorphic
aligned
magma
igneous
basalt
granite
obsidian
pumice
lava
erupts
volcanoes
volcanic ash
gabbro

8H.1 How sedimentary rocks form

In Unit 8G, you learned how James Hutton (1726–97) taught us that looking at what is happening to rocks now helps us to understand what happened to rocks in the past.

You looked at how:

- the weather breaks down rocks;

- rivers transport the rock fragments to the sea;

- layers of **sediment** are laid down on top of one another on the sea bed.

All these things are happening now. They have happened for over 4000 million years.

Rocks made from sediments are called **sedimentary rocks**.

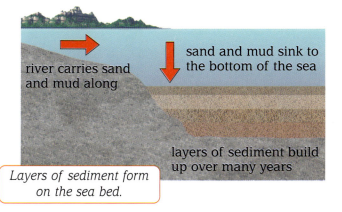

river carries sand and mud along

sand and mud sink to the bottom of the sea

layers of sediment build up over many years

Layers of sediment form on the sea bed.

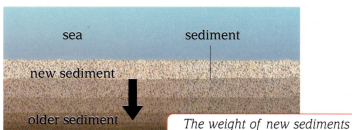

sea

sediment

new sediment

older sediment

The weight of new sediments presses down on the older sediments. This pressure squeezes water out and compresses the sediments. Chemical changes cement the fragments together, forming solid rock.

1 What are sedimentary rocks?
2 Write a paragraph about how sediments change into sedimentary rocks.

Millstone grit

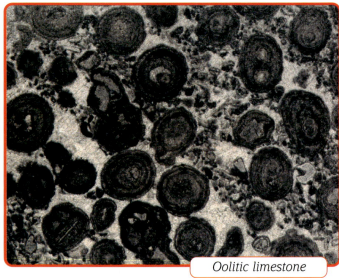

Oolitic limestone

3 Look at the pictures of the two rocks. Write down:

a <u>one</u> difference between them;

b <u>three</u> similarities between them.

You can recognise sedimentary rocks in two main ways:

● by their texture;

● by the presence of **fossils**.

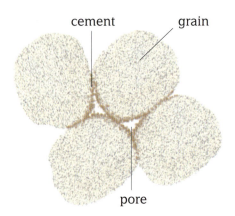

cement   grain

pore

crystals

The grains of sedimentary rocks do not interlock. There are often pores or spaces between the grains. We say that sedimentary rocks are porous

The grains of other rocks interlock like the pieces of a jigsaw. There are no pores, so they are non-porous.

Fossils are the remains of living things. Normally, you only find them in sedimentary rocks.

4 Draw a diagram to show the texture of sedimentary rock. Label the grains and the pores.

8H.2 Limestones are not all the same

Limestones form in the sea, so they are sedimentary rocks. You learned in Unit 8G that limestones often form from shells, corals and other hard parts of living things.

Some different limestones.

1 Look at the picture of different limestones.

   a Describe limestone A.

   b Write down a difference between limestone A and the other two limestones.

Limestones are carbonate rocks. They usually form in clear seas with very little sand or mud. Remember that you can test for carbonates using hydrochloric acid.

Testing limestone A. The new substances produced in this reaction are carbon dioxide, calcium chloride and water.

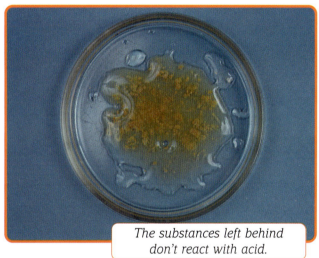

The substances left behind don't react with acid.

2 Look at the pictures.

   a What is in the bubbles?

   b Write a word equation for the reaction between limestone and hydrochloric acid.

3 **a** What is left behind when the reaction is finished?

   b What can you say about these substances?

4 Look back at the pictures of the three limestones. You have already tested limestone A. Now imagine you are testing limestone B.

   a What will be the same about the two test results?

   b What will be different?

   c Explain your answers.

8H.3 Rocks are sometimes changed

Sometimes new rocks form when existing rocks are changed without being melted. We call the new rocks **metamorphic** rocks.

- If you see a word with 'meta' in, it means change.
- 'Morph' in a word means form.
- So a 'metamorphic' rock is a rock with a changed form.

We call the process **metamorphism**.

limestone → marble

> You can see rounded grains and pores in limestone. Marble is harder, with a granular, sugary texture and no pore spaces.

shale → slate

> Shale is soft and crumbly, but slate is hard. The minerals in slate are lined up in the same direction.

sandstone → quartzite

> Sandstone is made up of grains of sand. Quartzite is harder with a sugary texture.

1 What is a metamorphic rock?
2 Write down the names of <u>three</u> metamorphic rocks.

3 Are metamorphic rocks porous or non-porous?
Explain your answer.

4 Look at the photographs on page 90. Write down
<u>three</u> types of changes that happen as a result of
metamorphism.

What causes the changes?

Heat or pressure, or a mixture of the two, can change rocks.

● <u>Pressure</u> lines up thin, flat, plate-like minerals in the same
direction. We say that they have become **aligned**.

● <u>High temperatures</u> cause reactions that chemically change
some of the minerals in the rocks.

All these changes happen in the solid state. The rocks do not melt.

*As you go deeper under the Earth's surface, the
pressure and temperature increase. This South
African gold mine is so hot that the miners can
only work for a few hours at a time.*

*This molten lava heats up the
surrounding rocks.*

5 Write down <u>two</u> ways that rocks get hot enough to
change them.

6 Write down <u>two</u> ways that pressure is put on rocks in
the Earth.

*Movements of the Earth's
crust cause heating and
squashing of rocks.*

7 What do you think happened to change the shale
into slate?

8H.4 Rocks formed from molten magma

When you heat ice, it melts to form water. When you cool the water to 0 °C, it solidifies again. We call 0 °C the melting point of ice. It is also the freezing point of water.

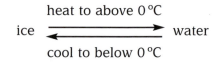

$$\text{ice} \underset{\text{cool to below 0 °C}}{\overset{\text{heat to above 0 °C}}{\rightleftarrows}} \text{water}$$

Deep in the Earth, the rocks sometimes get so hot that they reach their melting point. We call molten rock **magma**. For example, the melting point of granite is about 1000 °C. When magma cools to below 1000 °C, it solidifies, but it is still very hot. Other rocks melt at different temperatures.

1 Draw a diagram, like the one for ice, to show what happens to granite above and below 1000 °C.

a sandstone

Magma is a mixture of minerals. When magma solidifies, it usually forms crystals. You can see crystals of different minerals in most rocks formed from magma. We call rocks formed from magma **igneous** rocks.

2 What is:

a magma;

b igneous rock?

b basalt

3 Look at the photographs of the three rocks.
Which <u>two</u> are igneous rocks? Explain your answer.

4 Write down <u>one</u> difference between the two igneous rocks.

c granite

Crystal size in igneous rocks

In Unit 7E, you made crystals of salt by evaporating water from a solution.

5 a What does 'evaporating' mean?

b What affects the size of the crystals formed when water evaporates from a solution?

c How do you make large crystals?

these crystals formed quickly

these crystals formed slowly

The size of the crystals depends on how quickly you evaporate the water from the solution.

As magma solidifies, it is the rate of cooling that affects crystal size. When magma cools slowly, the crystals have a long time to form. So the crystals are larger than when magma cools quickly. Sometimes magma cools so quickly that there is no time at all for crystals to form.

6 Look back at the photographs of granite and basalt. Did granite or basalt form from magma that cooled quickly? Explain your answer.

Some magmas cool more quickly than others

Molten magma rises up through the Earth's crust.

Obsidian *is glassy, not crystalline, so it is sometimes called volcanic glass.*

- Some magmas cool and solidify while they are still deep underground. They are surrounded by hot rock, so they cool slowly.

- Some magmas solidify in cracks in rocks or between layers of other rocks nearer to the Earth's surface. They cool more quickly.

- Others magmas are still molten when they come out onto the surface. They cool very quickly.

7 Look at the pictures of obsidian and pumice. Describe how and where each of them probably formed.

8 Look back at the photographs of granite and basalt. Which one formed deep underground? Explain your answer.

You can see gas bubbles trapped in **pumice**, *but no crystals.*

We call magma that reaches the surface of the Earth **lava**. When lava comes out onto the surface of the Earth we say that it **erupts**. Some lavas are very runny so they form lava flows that spread out in layers over large areas.

Lava flows out of a volcano called Kilauea, in Hawaii

Less runny lavas form **volcanoes**. Sometimes magma solidifies inside the volcano. Gases are trapped underneath the solid magma. The pressure builds up until the gases, solids and more molten lava burst out of the top like the froth from a shaken fizzy drink. These materials form a cloud which settles as a **volcanic ash** cone. Eruptions can happen on land, into the sea or even under the sea.

Mount St. Helens is an ash cone. Ash is blown out of this volcano in a cloud.

9 Describe the shape of a volcano.

10 Write down <u>two</u> substances that volcanoes are made from.

11 Describe how the volcano in the drawing was formed.

12 Surtsey first appeared as a cloud of steam in the sea. Explain this as fully as you can.

13 Find out when Surtsey appeared out of the sea. Then use an atlas to find out where it is.

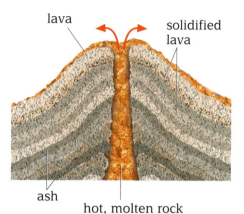

Some volcanic cones form from ash and lava.

The minerals in magmas vary too

Mineral	Percentage composition	
	Gabbro	Granite
silica	49.9	70.8
aluminium oxide	16.0	14.6
iron and magnesium minerals	18.2	4.3
other minerals	15.9	10.3

Surtsey erupted under the sea and formed a new island.

Gabbro

Granite

14 Write down <u>one</u> piece of evidence that both rocks:

a are igneous rocks;

b formed deep below the Earth's surface.

8H.5 Recycling rocks

In this unit you have learned that you can classify rocks as sedimentary, igneous and metamorphic.

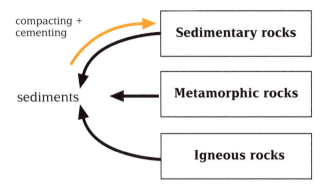

Sediments are deposits of rock fragments. They form as a result of weathering, erosion and deposition. Sedimentary rocks form by compacting and cementing of the sediments.

1   Which colour of arrow on the diagram shows

   a weathering, erosion and deposition;

   b compacting and cementing?

Magma is molten rock. Igneous rocks form when magma cools and solidifies.

Rocks changed by heat and pressure are called metamorphic rocks.

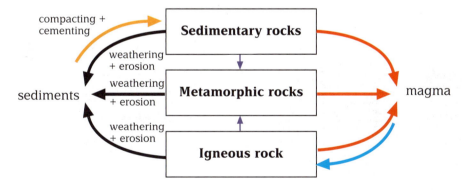

2   What colour are the arrows that show:

   a melting;

   b cooling and solidifying;

   c heat and pressure?

So, the minerals in rocks are used over and over again for <u>millions of years</u>. In other words, they are recycled.

3   What name do we give to the recycling of rocks?

You should now understand the key words and key ideas shown below.

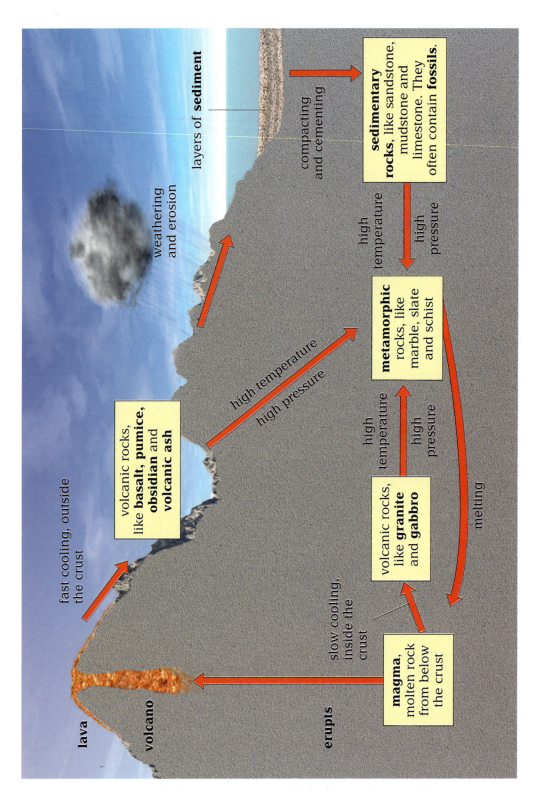

Heating and cooling

In this unit we shall study heating and cooling and learn how substances behave when they heat up and cool down. We also learn about thermometers and how heat energy is transferred from one place to another.

KEY WORDS

temperature
thermometer
Celsius
degrees
heat energy
joule
conduction
thermal insulators
convection
radiation
melting point
evaporation
boiling point
condensation

8I.1 Measuring how hot things are

PROFESSOR ASSAM IS FLYING IN TO LONDON FROM INDIA.

HIS OLD FRIEND DOCTOR CHILBLAIN, AN EXPERT IN LOW TEMPERATURE PHYSICS, LEAVES HIS LAB IN ICELAND ABOUT THE SAME TIME.

THEY ARE BOTH MET AT HEATHROW BY HIS COUSIN INDIRA WHO LIVES IN LONDON.

PROFESSOR ASSAM FEELS COLD. DOCTOR CHILBLAIN FEELS TOO HOT. INDIRA THINKS THEY ARE BOTH WRONG AFTER ALL IT IS JUNE AND VERY PLEASANT WEATHER IN LONDON.

People judge how hot or cold things are by the feelings they get from their skin. There is a problem with this. Your estimate depends on where your skin has just been! If you move from a hot place to a cool place you will think it is colder than it actually is. How hot or cold something is is called **temperature**. To get an accurate temperature you need something that always gives the same answer no matter where it has just been. Your skin is no good for this! We use an instrument called a **thermometer**.

There are many different types of thermometers. Some work by electricity, some use a substance that changes colour, others use a liquid that expands and contracts when the temperature changes. The differences do not matter because they all give a reliable measurement of the temperature.

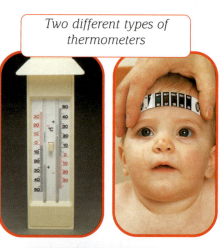

Two different types of thermometers

1 What is the name of the instrument used to measure temperature?

2 What does the temperature of something tell us?

3 Explain why using your skin to estimate temperature can be unreliable.

The Celsius scale

Over the years, many temperature scales have been used. One of the first scientists to try measuring temperature was Galileo. He invented the first thermometer in 1600. It was not very good by today's standards but it was a start. It was based on the expansion of air.

Over the next 150 years several different scientists worked on improving Galileo's ideas. In 1742, Anders Celsius, a Swedish scientist, invented a simple temperature scale with 100 divisions. The 0 mark and the 100 mark could be found by using melting ice and boiling water, so it was easy for anyone to make a thermometer.

This scale is called the **Celsius** scale in honour of Anders Celsius. A temperature of twenty **degrees** Celsius is written as 20°C.

Galileo's thermometer

4 What is the name of the scale used for measuring temperature in everyday work in science?

5 How should you write a temperature of 30 degrees Celsius?

6 What is the temperature range in Britain, from a cold winter's day to a hot summer's day?

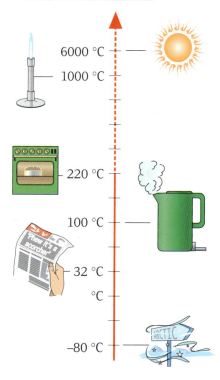

6000 °C

1000 °C

220 °C

100 °C

32 °C

°C

−80 °C

8I.2 Changing the temperature

Heat energy and temperature are different things. This section looks at the differences between them.

Temperature tells us how hot something is. A boiling kettle contains water at a temperature of 100 °C. We say the temperature of the water goes up because it has been provided with **heat energy**. Even though the water in a swimming pool is at a lower temperature than the water in a kettle, the swimming pool contains much more heat energy because it contains much more water. Like all types of energy, heat energy is measured in units called **joules**. The symbol for the joule is J, so 100 joules is written as 100 J.

1 What do you supply to something to make its temperature go up?

2 Find out who James Prescott Joule was and why the unit for energy was named after him.

From high to low

Heat energy moves from high temperature areas to low temperature areas. Something at a high temperature can transfer heat energy to something that is cold. The temperature of the cold object will rise. The temperature of the hot object will fall unless it has a supply of energy to keep its temperature up. For example, if a house is heated so that the temperature inside is higher than outside, heat energy will move out through the walls and windows. Also, if you open a fridge, heat energy moves from the room into the fridge because the room is at a higher temperature. We <u>never</u> say that 'cold moves'.

3 What can move from a hot object to a cold object?

4 What will happen to the temperature of a cold object if it is supplied with heat energy from a hot object?

5 If you put a piece of hot metal into a cold drink, the drink will warm up. Explain why this happens and what happens to the metal.

81.3 Hot to cold

Heat energy spreads out from hot places to cold places. It does this in a special way in solids.

Conducting the heat energy

Imagine a block of solid material with one end hot and the other end cold.

The left-hand side is heated with a Bunsen burner. The particles on the left vibrate quickly because that end of the block has a high temperature. Because the particles in a solid are close together, movement gets passed along to the particles on the right. The temperature on the right will start to rise as heat passes energy along the bar.

We say that heat energy has been transferred along the bar. This is called **conduction**. It is the only way heat travels through solids.

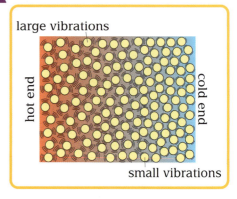

large vibrations

hot end

cold end

small vibrations

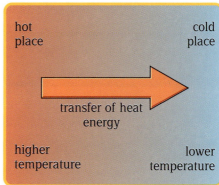

hot place

cold place

transfer of heat energy

higher temperature

lower temperature

1   When heat passes along a solid, what is passed from particle to particle inside the solid?

2   What is the name for the process that takes place when heat travels along a solid, passing from particle to particle?

Some solids conduct heat energy better than others

Sally and Harry have just made some tea. Harry has poured his into a metal mug. Sally has chosen a ceramic mug. Harry learns about conduction the hard way!

Metals are good conductors of heat. The particle vibrations move through metals easily. Other substances like wood, plastic, clay and cloth are poor conductors of heat. We call them heat insulators. They are also called **thermal insulators**.

Even though all metals conduct heat well, some metals are better at it than others. Copper is one of the best. Really expensive cooking pans often have copper in the base to help the transfer of heat energy from the flame to the food.

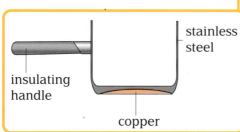

insulating handle

stainless steel

copper

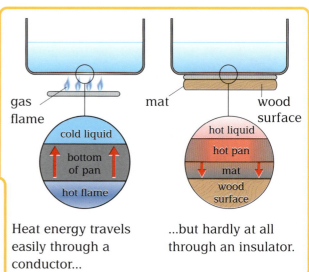

gas flame

mat

wood surface

cold liquid

bottom of pan

hot flame

hot liquid

hot pan

mat

wood surface

Heat energy travels easily through a conductor...

...but hardly at all through an insulator.

Aluminium is another good conductor. You can use it to speed up the cooking of baked potatoes in the oven.

3   Which is the better conductor of heat, metal or plastic?

4   Explain why pans used for cooking often have a metal base but a plastic handle.

5   Name two metals that are good heat conductors.

6   Suggest a reason why joints of meat with a bone in can take less time to roast than the same meat that has had the bone taken out.

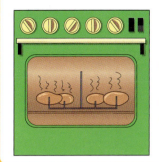

It takes a long time for baked potatoes to cook through to the middle

Using aluminium spikes, they cook through in half the time.

Air is a very poor conductor

The particles in a gas are spread out. They fly about and hit each other occasionally, but most of the time they are not connected. Air is a gas. This makes air a very poor conductor because the particles cannot pass the heat energy on easily.

We use air as a good insulator in lots of ways. Lots of layers of clothes keep you warmer in winter than one thick garment. This is because each layer traps some air. The layer of air acts as an insulator. Polystyrene cups made to keep drinks hot have tiny bubbles of air trapped in their walls.

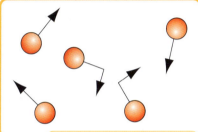

The particles in a gas move about and collide with each other. There are far fewer particles in each cubic centimetre than there are in a solid or a liquid. So gases are poor thermal conductors.

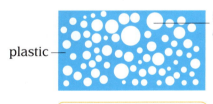

plastic

small bubble of trapped air

Foam plastic insulation

fibre

small 'pocket' of trapped air

Clothing fibre insulation

7   How are the particles in a gas arranged?

8   What is it about the arrangement of particles in a gas that makes it a poor conductor of heat?

9   How do clothes make use of air to keep the heat in?

Conduction makes some things feel cold

Sometimes, two things at the same temperature feel like they are at different temperatures when you touch them. An example of this happens when you touch the handlebars and the seat of a bike.

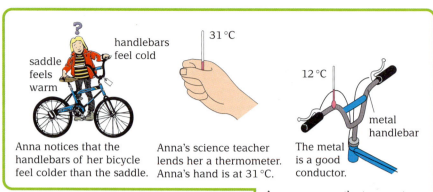

Anna notices that the handlebars of her bicycle feel colder than the saddle.

Anna's science teacher lends her a thermometer. Anna's hand is at 31 °C.

The metal is a good conductor.

This happens because of conduction. The metal of the handlebars is a good conductor. It is colder than your hand so heat energy flows quickly out of your hand when you touch the metal. The flow of heat energy makes your hand feel cold. The plastic seat is not such a good conductor so it feels a lot warmer when you touch it. This is because the heat energy does not flow out of your hand so quickly.

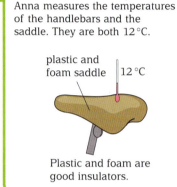

Anna measures the temperatures of the handlebars and the saddle. They are both 12 °C.

Plastic and foam are good insulators.

10 Which feels colder when you touch it, the handlebar or the saddle? Give a reason for your answer.

11 If you had to get a spade from the garden shed on a frosty day, would it be better to get hold of the wooden part or the metal part? Give a reason for your answer.

8I.4 Moving the heat energy in liquids and gases

You can get heat energy to move through liquids and gases, but it happens in a different way to the conduction of heat in solids.

Boiling a kettle

Look at the pictures of a kettle.

- The heating element of a kettle heats the water around it.
- The particles in the hot water move faster and move further apart.
- The hot water is less dense than the cold water around it. So the hot water rises up through the cold water.
- Cold water replaces the hot water around the heating element and starts to heat up.
- The water cycles around the kettle until it boils.

This is called **convection**.

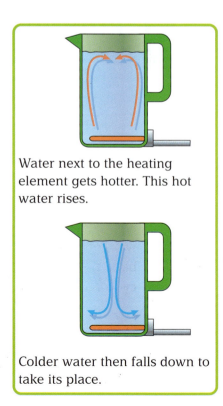

Water next to the heating element gets hotter. This hot water rises.

Colder water then falls down to take its place.

Gases and liquids are both fluids. So the same idea applies to the air in a room. Hot air is less dense than cold air, so it rises up from around the wall heater. The cold air moves in to replace it. The cycle of air is known as a <u>convection current</u>. Heat transfers in this way in liquids and gases – because they are fluids, they can flow.

You can see the convection currents in water by using a dye. You can see the effect of convection currents in air with a spiral made of thin paper. It turns as hot air rises past it.

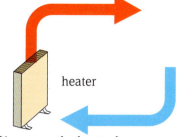

heater

Air next to the heater becomes hotter. This hot air rises. Colder air then falls to take its place.

1 Which is denser, hot water or cold water?

2 What is the name for the current that forms when hot water floats upwards and cold water sinks?

3 What happens to the particles in water when the temperature increases?

4 How does the increase in temperature of water affect the density of the water?

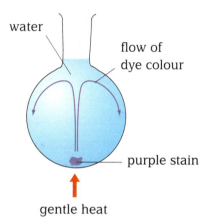

water

flow of dye colour

purple stain

gentle heat

More convection currents

Convection currents also occur in nature. Farm buildings, ploughed fields and tarmac warm the air so convection currents called thermals rise above them. Glider pilots use them to gain height.

5 What is the name for the convection currents used by glider pilots to gain height?

6 Describe how convection currents move inside a fridge.

Hint: the cooling unit is at the top.

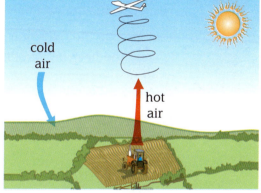

cold air

hot air

A lava lamp also uses convection. The bulb at the bottom of the lamp heats the coloured wax inside. The hot wax rises upwards because it is less dense than the liquid. The wax cools when it gets to the top of the lamp and sinks back down. It gets heated again at the bottom and the cycle continues.

7 Sea breezes often blow from the sea towards the land during the day. At night they often blow from the land towards the sea. Find out why this happens.

81.5 How energy travels through space

During the daytime, heat energy reaches the Earth's surface from the Sun. The energy has taken about 8 minutes to travel across 150 million kilometres of empty space. It does not happen by conduction or convection, because there is nothing between the Sun and the Earth to allow conduction or convection to happen. This method of heat transfer is called **radiation**. We say that the Sun radiates heat energy in the same way that it radiates light.

 1 How is the energy from the Sun carried to the Earth?

 2 Through what can heat transfer by radiation happen while heat transfer by conduction and convection cannot?

Reflecting heat radiation

A shiny surface reflects heat radiation just like it reflects light. You can feel this effect if you hold your hand in front of a spotlight bulb in a reading light and compare it to an ordinary bulb at the same distance.

The spotlight bulb has a shiny coating to reflect more light forwards. The shiny coating also reflects radiated heat energy.

You can use the reflection from a shiny or light-coloured surface to keep the heat out of some things. Houses in hot countries are often painted white to reflect the heat radiation and keep the building a bit cooler. Fridges and freezers are usually shiny white to reflect heat radiation. Petrol storage tanks at refineries usually have shiny walls to reflect the heat radiation from the Sun.

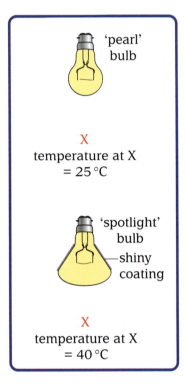

'pearl' bulb

X
temperature at X
= 25 °C

'spotlight' bulb
shiny coating

X
temperature at X
= 40 °C

 3 What can a shiny or light-coloured surface do to heat radiation?

4 Why are some objects coloured white or given shiny surfaces to keep them cool?

 5 How might spotlight bulbs be used in restaurants to keep plates of food hot until they are taken out and served?

8I.6 Keeping warm

Keeping the house warm

A three-bedroom house in Britain can cost as much as £15 a week to keep warm. You can reduce this by half if you control the heat loss to the outside air. This diagram shows how much energy you can lose every second from a house on a winter's day. The total number of joules shown on the diagram would be enough to bring 2 litres of cold water to the boil in about one minute!

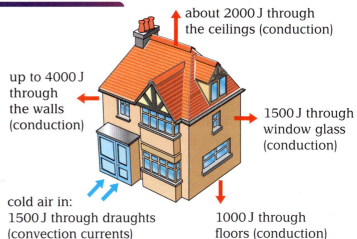

about 2000 J through the ceilings (conduction)

up to 4000 J through the walls (conduction)

1500 J through window glass (conduction)

cold air in: 1500 J through draughts (convection currents)

1000 J through floors (conduction)

One of the cheapest things you can do is to insulate your loft with layers of glass fibre. Convection currents cause the hot air to rise up inside your house. The glass fibre traps air, which acts as a good insulator and reduces heat loss by conduction though the roof. It costs about £300 to insulate a loft. In a year you could save about £150 on your heating bills.

Draught excluders only cost a few pounds to fit and they can save about £50 in a year. Double glazing is more expensive. It would probably take over ten years for someone to save enough to pay for it on their heating bills. Double glazing works by trapping a layer of air between two sheets of glass. The air acts as an insulator. Apart from reducing heat loss, it also reduces the level of sound coming in from outside. Many people have double glazing fitted because it makes it quieter inside the house.

The radiated heat from the back of a radiator can be reflected back into the room by a shiny plastic sheet on the wall. Sometimes the walls of a house are made of two layers of brick with plastic foam between them. This is known as cavity wall insulation.

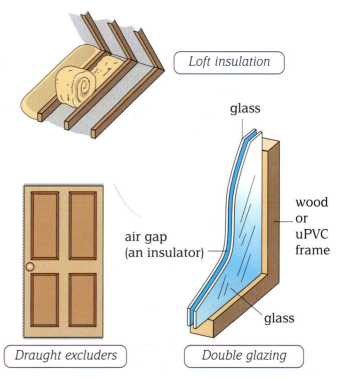

Loft insulation

glass

air gap (an insulator)

wood or uPVC frame

glass

Draught excluders

Double glazing

1 Describe <u>two</u> ways of reducing heat loss from your house. Explain how <u>one</u> of these works.

2 Why might someone have double glazing fitted if they only plan to live in the house for a few years? Suggest several different reasons.

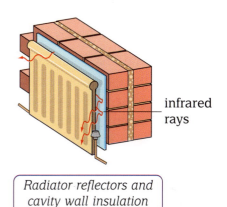

infrared rays

Radiator reflectors and cavity wall insulation

The Thermos flask

A Thermos flask keeps hot things hot and cold things cold. Its walls are designed to prevent heat energy passing through them.

There are no particles in a vacuum so conduction and convection cannot happen. The only way that heat energy can cross a vacuum is by radiation. A shiny surface is a good reflector of radiation. The wall reflects back the small amount that reaches it. This means that very little radiated heat crosses the vacuum and escapes.

The stopper is filled with plastic foam, which contains trapped air to make it a poor conductor. The stopper also stops convection currents carrying heat out of the top of the flask.

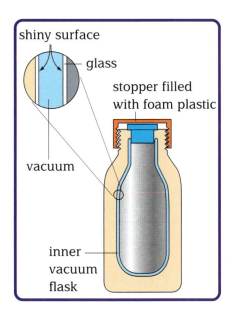

shiny surface

glass

stopper filled with foam plastic

vacuum

inner vacuum flask

3 What is a vacuum?

4 What is the only method of heat transfer that can cross a vacuum? Give a reason for your answer.

5 Find out who James Dewar was and what his connection to the Thermos flask is.

81.7 Changing between solid, liquid and gas

When you want to change a solid to a liquid or a liquid to a gas, you have to heat the solid or liquid up to get its particles moving faster. You also need to provide some extra energy so the particles can separate from each other.

Monitoring the temperature

A beaker containing 200 g of crushed ice is taken from a freezer at −15 °C. It is heated until the ice has melted and warmed up to about 60 °C. The temperature is logged at intervals of time as the ice is melting. A computer plots a graph of temperature against time.

The first section of the graph shows the temperature rise while the ice warms up. Once it reaches 0 °C the temperature stays the same while the ice melts. This happens because the heat energy being supplied is used to get the particles to break loose from each other rather than go any faster.

The water is changing from a solid to a liquid. The temperature where a solid changes into a liquid or a liquid changes into a solid is called its **melting point**. Once all the ice has melted the heat energy starts to make the particles move faster and the temperature starts to rise again.

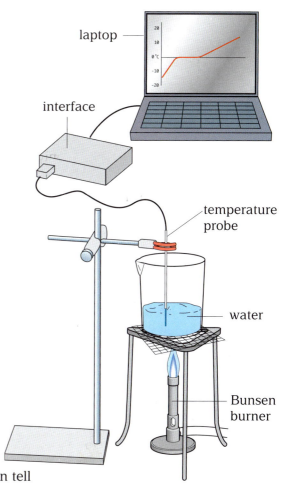

laptop

interface

temperature probe

water

Bunsen burner

1 What is happening to the ice where the temperature stays the same on the graph?

2 What does the energy supplied to the particles in the melting ice do while the temperature stays the same?

3 What happens to the temperature of the water once all the ice has melted? Give a reason for your answer in terms of what is happening to the particles.

About melting points

Different substances have different melting points. You can tell where the melting point is by plotting a graph of the temperature as something melts. The melting point is the temperature where the graph flattens out while the particles break free of each other. You can also find the melting point from a <u>cooling curve</u>. A cooling curve is the same type of graph in reverse, where you cool the liquid down into a solid and plot its temperature. This is the type of graph you get when a liquid cools and changes to a solid.

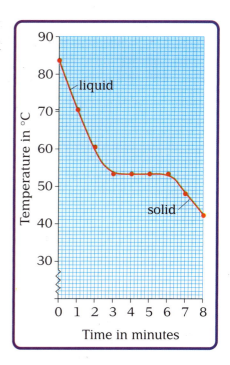

4 At what temperature does the liquid start changing to a solid?

5 How long does it take for the liquid to change into a solid, from the moment it starts to solidify?

6 How do you think the shape of the graph would change if the experiment was repeated in much colder surroundings?

Evaporation, boiling and condensation

As you heat up water, some of the faster particles start to escape from the surface. We call this **evaporation**. Evaporation happens at all temperatures. You can see this on a cold day after it has rained. Even though it takes a while, eventually the roads and pavements will dry out by evaporation. On a hot summer's day the roads and pavements dry out much more quickly. Sometimes you can even see steam in the air above them.

When you heat water so that bubbles of evaporated water form everywhere in the body of the water, we say the water is boiling. The temperature at which a liquid boils is called its **boiling point**.

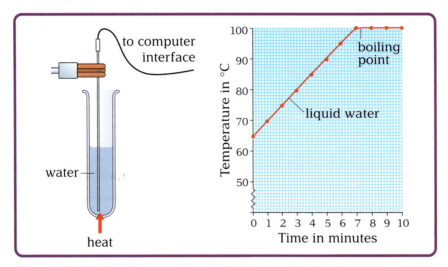

Once water boils, the heat energy being supplied to it goes into separating the particles to form a gas. This means steam at 100 °C has a lot more energy than water at 100 °C. A scald from boiling water is pretty bad but a scald from steam is a lot worse! If you cool water vapour or steam down it changes back to water. We call this **condensation**. The steam that emerges from a kettle or a steam iron is visible because it is beginning to condense.

7   What effect causes pavements to dry after it rains?

8   What is the name for the effect when bubbles of vapour form all the way through a body of water?

9   What is condensation?

10   Why can a scald from steam be worse than a scald from boiling water?

11   Find out who James Watt was. What was his connection with steam?

You should now understand the key words and key ideas shown below.

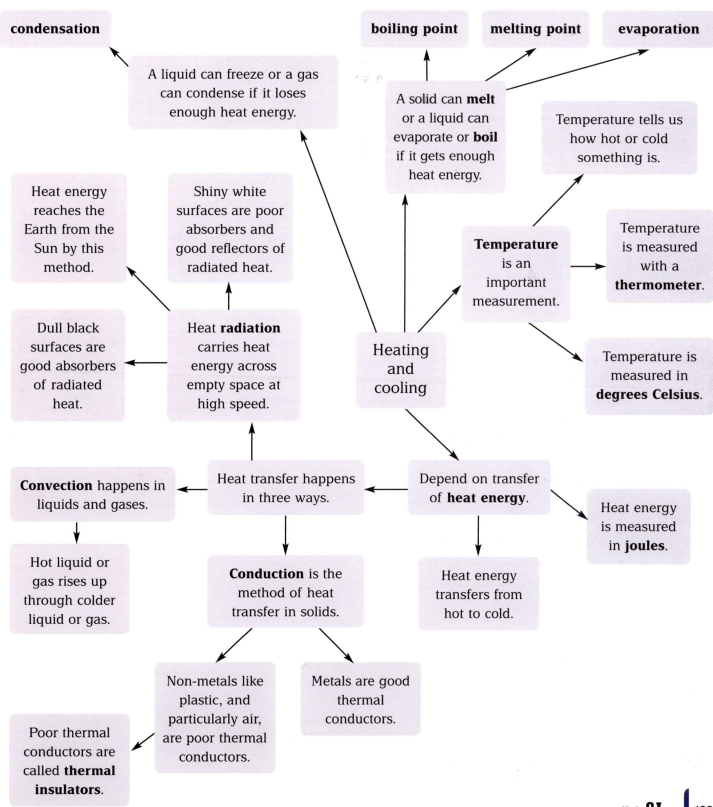

condensation

A liquid can freeze or a gas can condense if it loses enough heat energy.

boiling point **melting point** **evaporation**

A solid can **melt** or a liquid can evaporate or **boil** if it gets enough heat energy.

Temperature tells us how hot or cold something is.

Heat energy reaches the Earth from the Sun by this method.

Shiny white surfaces are poor absorbers and good reflectors of radiated heat.

Temperature is an important measurement.

Temperature is measured with a **thermometer**.

Dull black surfaces are good absorbers of radiated heat.

Heat **radiation** carries heat energy across empty space at high speed.

Temperature is measured in **degrees Celsius**.

Heating and cooling

Convection happens in liquids and gases.

Heat transfer happens in three ways.

Depend on transfer of **heat energy**.

Heat energy is measured in **joules**.

Hot liquid or gas rises up through colder liquid or gas.

Conduction is the method of heat transfer in solids.

Heat energy transfers from hot to cold.

Poor thermal conductors are called **thermal insulators**.

Non-metals like plastic, and particularly air, are poor thermal conductors.

Metals are good thermal conductors.

Magnets and electromagnets

In this unit we shall be studying magnets and electromagnets, what they can do and how to make them, and familiar objects that use them. We shall learn about magnetic fields and how to plot them.

8J.1 Magnetic forces

A **magnetic force** exists between two magnets and between a magnet and any magnetic material. This can be a push or a pull.

Different types of materials

Many years ago, the Ancient Greeks discovered a rock called lodestone. This rock could pull another lump of lodestone towards it. They **attract** each other. If it was turned around it could push the other lump of lodestone away. They **repel** each other. Lodestones in the ground also attracted some other substances. They made shepherds' shoes stick to the ground by attracting the iron tacks in their soles. The Greeks found that they could separate materials into three groups:

● **Magnets** These most commonly contain iron and attract and repel each other. They also attract magnetic materials.

● **Magnetic materials** These include **iron**, **steel**, **nickel** and cobalt. They are attracted to magnets but not to each other.

● **Non-magnetic materials** These are not affected at all by magnets. Non-metals and some metals, such as copper and aluminium, are examples of non-magnetic materials.

1 What <u>three</u> groups did the Greeks put materials into?
2 What happens to magnetic materials near magnets?

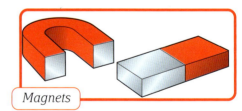

Magnets

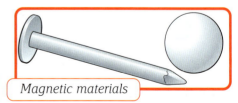

Magnetic materials

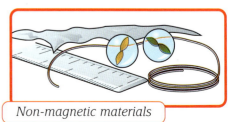

Non-magnetic materials

Magnetic attraction

Magnets only attract magnets and magnetic materials. Steel is a magnetic material but aluminium is not, so magnets are used to sort aluminium cans from steel cans for recycling. The steel cans are attracted to the magnet, leaving the non-magnetic aluminium cans behind.

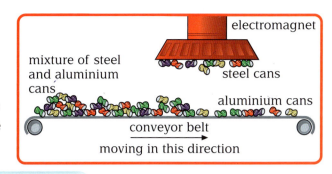

4 a Are steel cans magnetic or non-magnetic?

b Are aluminium cans magnetic or non-magnetic?

5 When the magnet is used, explain whether:

a the steel cans are attracted to it or left behind;

b the aluminium cans are attracted to it or left behind.

What magnetic forces pass through

Fridge magnets can hold pieces of paper or thin plastic to a painted fridge door.

A strong magnet held one side of your hand can move a magnetic object on the other side. The magnetic force passes through non-magnetic materials like paper, plastic, paint, skin and bone.

Dangle a paperclip near a magnet and the magnet will attract it. Now put a sheet of iron in the gap between the magnet and the paperclip. The magnet does not attract the paperclip as much because the iron partially stops the magnetic force passing through. Magnetic materials do not let magnetic forces pass through.

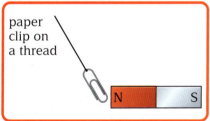

6 What type of material:

a lets magnetic forces pass through it;

b will not let magnetic forces pass through it?

7 Would magnetic forces pass through:

a aluminium;   **b** steel?

8 How would each of these materials behave near a magnet?

a nickel   **b** plastic   **c** iron

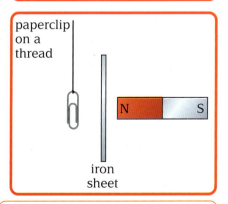

The iron sheet stops the magnet from attracting the paperclip.

Poles of a magnet

Early explorers discovered that when a magnet could move freely, it always came to rest with one end pointing towards the Earth's north. This is called the **north-seeking pole** or north pole for short. The other end point towards the Earth's south. This is called the **south-seeking pole** or south sole for short. The explorers used this to make a compass, which contains a magnet that is free to spin, to find their way in unknown places.

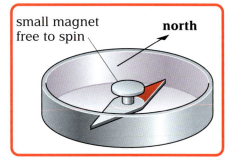

small magnet free to spin

north

8   What are the names given to the two ends of a magnet?

9   What happens when a magnet is left to move freely?

How the poles of a magnet behave

Magnets attract or repel other magnets. Some toy trains use magnets to hold the carriages together. Put the carriages facing the wrong way and they push each other apart. Two south poles, or two north poles will repel each other. However, magnets attract each other if a north pole and a south pole face each other. The magnetic force is strongest at the poles of a magnet.

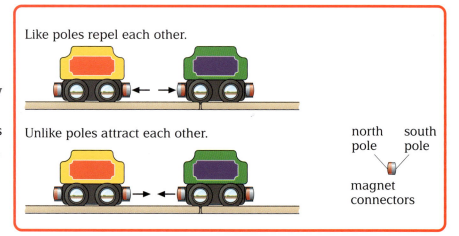

Like poles repel each other.

Unlike poles attract each other.

north pole   south pole

magnet connectors

10 Are these magnets attracting or repelling each other?

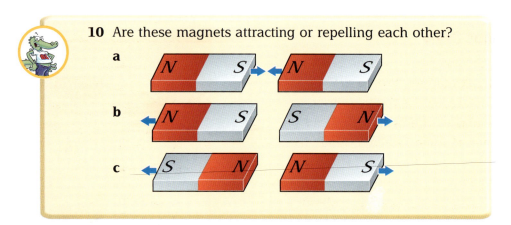

a

b

c

11 Does a compass always tell you which way is north? Find out about places where the needle of a compass does not point to north and explain why.

8J.2 Magnetic fields

There is an area around any magnet where that magnet would have an effect on a magnetic material. We call this area the **magnetic field**.

The shape of magnetic fields

Iron filings around a magnet arrange themselves into a pattern which shows the shape of the magnetic field around it.

You can use compasses to plot the direction of the magnetic field around a magnet.

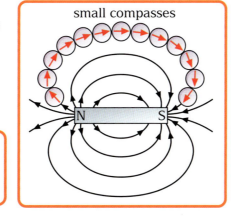

small compasses

1 What can you use to show someone where the magnetic field is around a magnet?

2 Draw a diagram to show the shape of the magnetic field around a bar magnet.

3 Use your diagram to show the direction of the magnetic field.

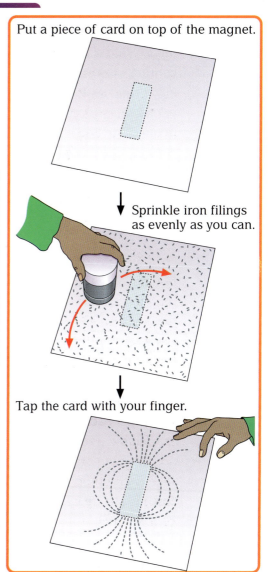

Put a piece of card on top of the magnet.

Sprinkle iron filings as evenly as you can.

Tap the card with your finger.

The strength of magnetic fields

The two diagrams show the magnetic field around a strong magnet and a weak magnet. Where the magnetic force is stronger, the field is drawn with more lines of force. Although the fields are the same shape, there are more lines of force drawn around the stronger magnet.

4 Where is the magnetic field of a magnet strongest?

5 What can you say about the number of lines of force for each magnet in the diagrams?

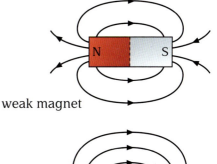

weak magnet

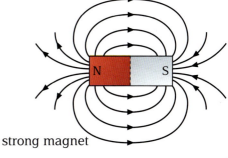

strong magnet

The Earth's magnetic field

A compass is used to help you find your way. This works because the Earth behaves as if it had a giant magnet inside it. This means that the Earth has a magnetic field. The magnet in the compass lines up with the Earth's magnetic field so that you can always tell which direction you are facing. In 1600, William Gilbert published work that described the Earth as a great spherical magnet. He had carried out experiments to back up his claim. He was a physician who worked for Queen Elizabeth I.

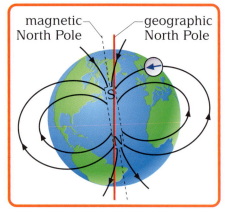

 6 What do you use a compass for?

7 What do we call the area around a magnet?

 8 How can you tell that the Earth has a magnetic field?

 9 If you placed a compass between two magnets, how could you tell which magnet was the stronger?

*The imaginary magnet which could produce the Earth's **magnetic field lines**. Its south pole points towards the Arctic and its north pole points towards the Antarctic.*

8J.3 Making and using magnets

You can make your own magnet using a magnet and an iron nail. Stroke the iron nail several times with one end of the magnet, and the iron becomes a magnet too. The iron will not stay magnetic for ever – it will lose its magnetism gradually. You can make temporary magnets in this way using any of the magnetic materials.

 1 What do you need to make a magnet?

 2 Name <u>three</u> materials that you can turn into a magnet by stroking them with another magnet.

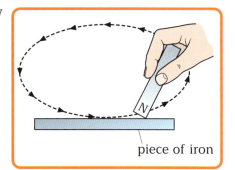

piece of iron

Using magnets

Magnets can be used to keep cupboard and fridge doors closed. A fridge door has a magnetic strip right around the door which keeps the door tightly closed.

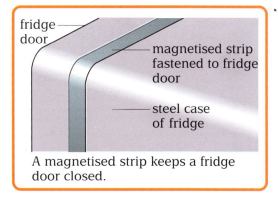

fridge door

magnetised strip fastened to fridge door

steel case of fridge

A magnetised strip keeps a fridge door closed.

Magnetic trains use magnets to float the train above the rail. This reduces friction and so the train can travel faster. The first design used magnets on the bottom of the train, which were repelled by magnetic forces in the rail.

3 How can you arrange two magnets so that they repel each other instead of attracting each other? Give <u>two</u> methods.

The diagram shows another design. The rain's magnets hang below the rail and are attracted upwards, lifting the train up to 5cm above the rail.

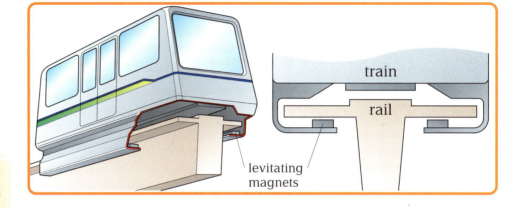

4 Suggest a material that the rails could be made from.

Magnetic tape contains many tiny magnetic particles. Information is stored on the tape when the particles are magnetised into different patterns. Because there are so many particles, the amount of information that can be stored is huge. Tape cassettes contain a length of magnetic tape, which is magnetised when a recording is made. This stores the information so the sound can be played back later.

When the information is recorded on the magnetic strip, the magnetic particles line up into a pattern.

Credit cards also have a magnetic strip, which stores information. Any magnets near to the magnetic strip can rearrange the particles, changing the information held.

Computers store information on floppy disks when the iron particles on the disks are magnetised into different patterns.

5 How does magnetic tape store information?

6 Why do you think it is important to keep credit cards and floppy disks away from strong magnets?

7 Stroking with a magnet is not the only way to turn a magnetic material into a magnet. Research another method.

8J.4 Making and using electromagnets

When an electric current flows through a wire, the wire behaves like a weak magnet. The wire can be made from any type of metal. It doesn't have to be a magnetic material. You can use compasses around the wire to show its magnetic field. Magnets made using electricity are called **electromagnets**.

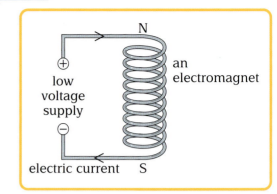

Electromagnets are very useful because they can be switched on and off. If you turn the current off, the electromagnet loses its magnetism. Switch the current on again and the magnet is switched back on.

1   How can you make an electromagnet?

2   What can you use to show the magnetic field around an electromagnet?

How to make electromagnets stronger

Some electromagnets are so weak that they cannot even pick up a paperclip. Others are strong enough to pick up a car! You can change the strength of the magnet by changing the size of the current – a larger current creates a stronger electromagnet. To make an even stronger electromagnet, you need to wind the wire into a **coil** and place a piece of iron inside the coil of wire. This piece of iron is called the **core**. It is important to choose the right material to make an electromagnet, however. Some magnetic materials, like iron, lose their magnetism when the current is switched off. These are called <u>soft</u> magnetic materials. Steel stays magnetised. It is called a <u>hard</u> magnetic material.

3   What are <u>three</u> ways to make an electromagnet stronger?

4   Give <u>one</u> useful difference between electromagnets and permanent magnets.

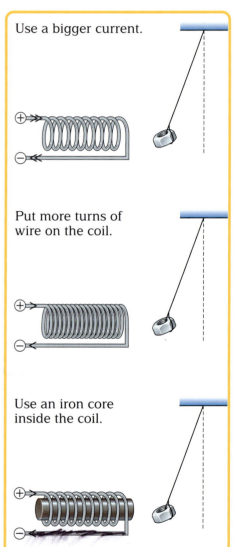

Ways to make an electromagnet stronger.

How do electromagnets work?

Wires carrying an electric current produce a magnetic field around them. If the wire is coiled, the electromagnetic field pattern is similar to that of a bar magnet. If an iron core is then placed inside the coil, the iron is magnetised. This is why the strength of an electromagnet is increased when an iron core is used.

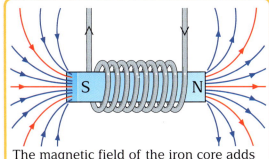

The magnetic field of the iron core adds itself to the magnetic field of the coil.

5 Why is an electromagnet stronger if it has an iron core?

6 What other materials would be good for making the core, so that the electromagnet is strong?

7 Find out what magnetic domains are.

Using electromagnets

Cranes in scrapyards use electromagnets to move old cars around. Switching the current on creates an electromagnet that is strong enough to lift the cars, which contain iron and steel. When a car has been moved to the correct place, the current is switched off. This turns off the electromagnet and the car is left in its new position.

electromagnet

The electromagnet in the crane lifts a scrap car.

When the crane driver switches off the current, the car falls.

8 Why are the cars attracted to the electromagnet?

9 How is the electromagnet switched on and off in a crane?

Some doorbells use electromagnets. When a switch is pressed, the circuit is complete and a current flows through the electromagnet. The electromagnet attracts a springy metal strip holding the hammer. When the hammer moves and hits the gong, it breaks the circuit. The current stops flowing, the electromagnet turns off and the metal strip moves back, completing the circuit. This starts the cycle once again.

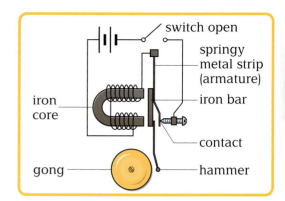

switch open

springy metal strip (armature)

iron core

iron bar

contact

gong

hammer

10 In a doorbell, what is attracted to the electromagnet?

11 How is the electromagnet switched off?

Relays

Relays use the current from one circuit to control the current in another circuit. When the first circuit is switched on, an electromagnet becomes magnetised, attracting the iron switch (called an armature) in the other circuit. The movement of the armature completes the second circuit and this allows a current to flow in the second circuit. The diagram shows a relay circuit that controls a light by remote control.

Car starter motors use relays. A small current flows in one circuit when the key is turned. This turns on the starter motor circuit, which uses a very large current. This is a safe way to switch on a large current.

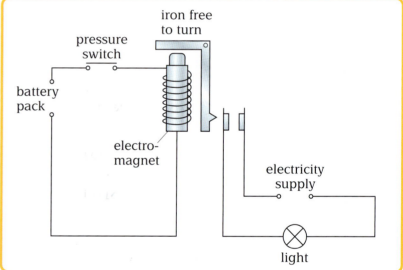

iron free to turn

pressure switch

battery pack

electro-magnet

electricity supply

light

This circuit controls a light that switches on if someone steps on the pressure switch.

12 Why is a relay switch used with a car starter motor?

13 Describe how the relay in the diagram switches on the light when the pressure switch is pressed.

You should now understand the key words and key ideas shown below.

Key words

magnetic force	magnet	magnetic material	non-magnetic material
attract	repel	nickel	steel
north-seeking pole	south-seeking pole	magnetic field	magnetic field line
electromagnets	core	coil	relay

Key ideas

Magnets create a **magnetic force** between one another and between themselves and magnetic materials.

Magnets can **attract** and **repel** one another. They also attract magnetic materials.

Magnetic materials are attracted to magnets but not to each other. Examples of magnetic materials are; iron, **steel** and **nickel**.

Non-magnetic materials are not affected at all by magnets.

The magnetic force passes through non-magnetic materials but not through magnetic materials.

The ends of magnets are called poles. The **south-seeking pole** points south and the **north-seeking pole** points north.

A north-seeking pole and a south-seeking pole attract each other, but two south-seeking poles or two north-seeking poles repel each other.

The area around magnets is called the **magnetic field**. **Magnetic field lines** can be plotted with a compass.

The direction of the magnetic field goes from the magnet's north pole to its south pole.

Stronger magnets are represented by showing more lines of force around them.

The Earth behaves as if it had a giant magnet inside it.

The magnet in a compass lines up with the Earth's magnetic field.

You can make your own magnet by stroking an iron nail with one end of a magnet.

When an electric current flows through a wire, the wire behaves like a weak magnet.

Magnets made using electricity are called **electromagnets**. These contain a **core**, usually of iron, and a **coil** of wire which carries an electric current.

Electromagnets are very useful because they can be switched on and off.

You can change the strength of an electromagnet by changing the size of the current, coiling the wire or placing a piece of iron inside the coil of wire.

Relay switches use the current from one circuit to switch on the current in another circuit.

A relay switch has an electro-magnet in one circuit next to an iron switch in the other circuit.

Light

In this unit we shall study light and learn how it behaves when it hits things and how colours are produced from white light. We shall also learn how light bounces off mirrors and how we can make light bend.

KEY WORDS
light source
ray
transmitted
reflected
absorbed
transparent
translucent
opaque
luminous
non-luminous
normal
image
object
refraction
prism
spectrum

8K.1 Where light comes from and how it travels

Light comes from sources

A **light source** is anything that produces light. The Sun is our main light source. There are places where light from the Sun does not reach, like windowless rooms or down mine shafts. Light from the Sun also fades towards the end of the day. Since prehistoric times humans have used other light sources to help them see when it is dark.

Until about 1900 most of our extra light sources involved burning something. Nearly all the extra light sources we use today, such as light bulbs, involve electricity. Electricity is more convenient. For example, you can store the energy for it in cells and batteries and move it from place to place along wires.

Electric light sources can be switched on and off instantly. Electric light sources are also safer than candles and oil lamps because they do not have flames in them, so there is less risk of starting a fire.

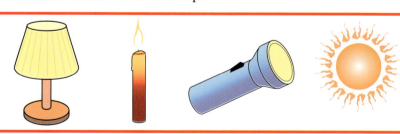

Examples of light sources

1   What is meant by a light source? Give <u>three</u> examples.

2   What is our main source of light?

3   Why is electricity more useful than candle wax or oil to power a light source? Give <u>four</u> reasons.

Light travels in straight lines

One of the earliest experiments with light looked like this.

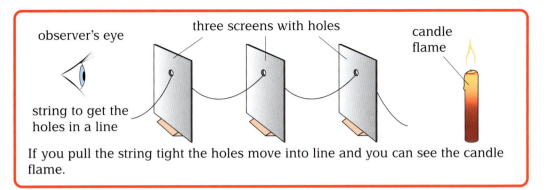

observer's eye

three screens with holes

candle flame

string to get the holes in a line

If you pull the string tight the holes move into line and you can see the candle flame.

The experiment shows that light travels in straight lines. Because of this we use straight lines with arrows to represent light. We call each line a **ray**. A light ray shows you the path that the light follows. A diagram that uses rays to show what light does is called a ray diagram.

The ray diagram shows how a torch and a pencil can be used to make a shadow. The light cannot go through the pencil. Because light travels in straight lines it cannot go round the pencil. That is why the pencil casts a shadow when it is put in front of the torch.

When you draw a ray diagram you just draw the main rays. You usually include the rays at the edges. You draw the rays as straight lines with arrows on them to show the direction the light travels in. The arrow always points away from the light source to show the direction.

The rays from the torch are spreading out. The light from the Sun is spreading out all over space. Because the Earth is so small compared to its distance from the Sun, the Sun's light rays reach us in a parallel beam.

This ray diagram shows how light rays from the Sun form a shadow of a tree.

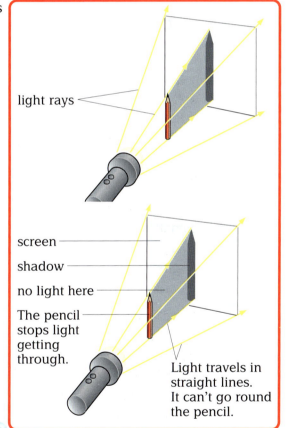

light rays

screen

shadow

no light here

The pencil stops light getting through.

Light travels in straight lines. It can't go round the pencil.

4 What is a light ray?

5 Why is a light ray drawn as a straight line?

6 Which property of light makes a pencil cast a shadow when it is held in front of a light source?

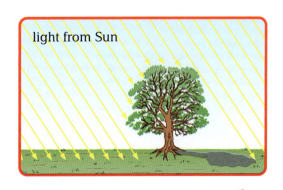

light from Sun

Light gets there straight away

If you go to watch a cricket match, you see the ball being hit but you hear the sound a short time later. This happens because light travels at an incredibly high speed and seems to reach you straight away. Sound travels about a million times slower and takes a fraction of a second to reach you. The first attempt to measure how fast light travels was by an Italian called Galileo Galilei in 1600. He tried uncovering a lantern and timing how long the light took to travel a few miles to

another person. His idea did not work because the smallest time he could estimate was about a tenth of a second. The time light actually took was only a few millionths of a second!

Scientists made the first really accurate measurements of the speed of light about 150 years ago using mirrors spinning very fast. These experiments

> Mama mia! I did not expect that! His light came on straight away. The speed of light is too quick for me to measure.

Distance	Time for light to travel that distance
From a light in the ceiling to the edge of a room in a house (about 3 m)	1 one hundred millionth of a second
From a lighthouse to a ship 20 miles away	1 ten thousandth of a second
From the Sun to the Earth (150 000 000 km)	500 seconds (about $8\frac{1}{2}$ minutes)
From the Sun to the edge of the Solar System (Pluto is about 5 900 000 000 km from the Sun)	About 20 000 seconds which is about $5\frac{1}{2}$ hours!
From the nearest star (Proxima Centauri) to us (about 40 400 000 000 000 km)	135 000 000 seconds, which is about $4\frac{1}{4}$ years!
From the edge of the observable universe to us (about 100 000 000 000 000 000 000 000 km)	About 10 000 million years!

were improved and changed, and we now know that light travels 300 000 000 metres in one second.

7 Why can you see someone hit a ball before you hear the sound produced when you are a few hundred metres away?

8 Why does the light seem to fill a room instantly when you switch an electric light on?

9 If the Sun suddenly went out one day, how long would it be until we noticed?

10 Find out what a light year is and why it is used to measure distances in space.

8K.2 Light hitting objects

When light hits something it can do one of three things:

- go through (be **transmitted**);
- bounce back (be **reflected**);
- stay inside and heat up the object (be **absorbed**).

Going straight through

Some substances let light go straight through them. Glass, water, air and some types of plastic are good examples. We say these substances are **transparent**.

1. What is a transparent substance?
2. Give two examples of transparent substances and where they might be used.

3. List some advantages and disadvantages of glass as a substance for windows.

Letting the light through but breaking it up

Sometimes you want to let the light through, but you do not want anyone to be able to see through. We use a **translucent** substance to break the light up. Examples in the home are:

frosted glass

cotton blind

Some light gets through, but you can't see through the blind.

- frosted glass;
- plastic or glass with very detailed patterns on;
- sheets of white cotton;
- 'pearl' light bulbs give an even glow compared with the glaring light you get from a clear bulb.

Clouds are translucent. The Sun's light is scattered when it comes through them. That is why there are no shadows on a cloudy day even though it is still light.

4. What is a translucent substance?
5. Give <u>three</u> examples of where a translucent substance might be useful.
6. What do clouds do to the Sun's light rays?

Glass is transparent (it lets light through).

Glass keeps wind and rain out.

Glass does not dissolve in rain.

Glass does not scratch or mark as easily as plastic.

If glass breaks, the pieces are very sharp.

A car windscreen is transparent.

The visor is transparent on a helmet.

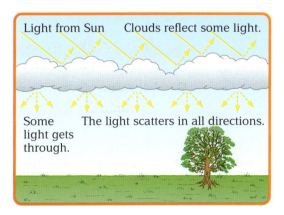

Light from Sun   Clouds reflect some light.

Some light gets through.   The light scatters in all directions.

Stopping the light

Some substances stop light going through them. We say that a substance that stops light is **opaque**. Metal and wood are opaque substances. Some types of plastic are opaque.

7 What is an opaque substance?

8 Give <u>two</u> examples of where an opaque substance might be useful.

film case

blackout blind

You must use an opaque substance for the case of a photographic film or the light will spoil it. The windows in a roof can be fitted with opaque blinds to shut sunlight out completely.

Bouncing off an opaque surface

Pale or white things reflect most of the light that falls on them. Black and dark things absorb most of the light that falls on them.

9 Opaque substances of which colours reflect most of the light that falls on them?

10 What can cyclists and pedestrians do to make sure that they are visible and safe at night?

White things and pale things reflect most of the light that falls on them.

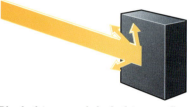

Black things and dark things reflect very little of the light that falls on them. They absorb most of the light.

Light entering your eye

You see an object when light from it enters your eye. For objects that give out light, like a light bulb, that usually means the light is travelling in a straight line from it to your eye. For objects that do not give out light, you see them when light reflects off them into your eye.

Objects that give out light are called **luminous** objects. Objects that do not give out light are called **non-luminous** objects.

11 What happens for you to see a luminous object?

12 What happens for you to see a non-luminous object?

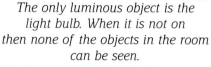

The only luminous object is the light bulb. When it is not on then none of the objects in the room can be seen.

13 Do you think that the Moon is a luminous or a non-luminous object? Give a reason for your answer.

14 What non-luminous objects can be seen from the Earth in the night sky? Can any of them be seen in the daytime? Give reasons for your answer.

8K.3 Mirrors

Mirrors reflect light without scattering it. This makes them very useful in many situations. A mirror reflects nearly all of the light that falls on it. The same is true of a piece of white paper, but you cannot see your face in a piece of paper.

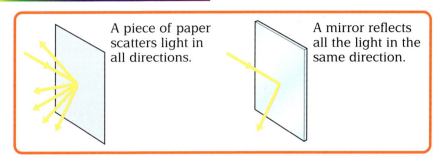

A piece of paper scatters light in all directions.

A mirror reflects all the light in the same direction.

A mirror reflects a ray of light at the same angle as it strikes it. We measure the angle of the light hitting and leaving the mirror from a reference line called the **normal**. The normal is a line drawn at 90° to the surface of the mirror.

The light hitting the mirror is called the incident ray. The light leaving the mirror is called the reflected ray. You can use mirrors to see things you could not normally look at.

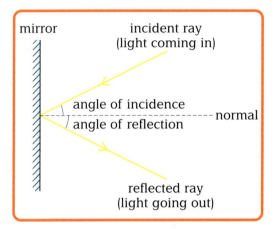

mirror

incident ray (light coming in)

angle of incidence

angle of reflection

normal

reflected ray (light going out)

Angle A and angle B are equal. The mirror reflects a beam at the same angle it strikes the mirror.

angle P

angle Q

The girl on the right can see the boy's hair because light bounces off the hair and enters her eye. The boy can see his own hair because light from it bounces off the mirror and then enters his eye.

Mirrors are sometimes used to help drivers 'see round corners'.

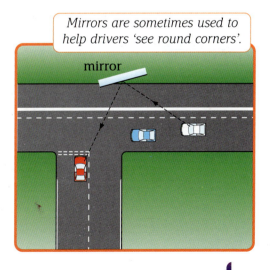

mirror

1   What is the difference between light bouncing off a piece of paper and light bouncing off a mirror?

2   What rule does light obey when it bounces off a mirror? Draw a diagram to show this.

3   Look at the road diagram. Draw a diagram to show where you would put a second mirror so that the driver of the blue car could see round the corner to the left.

Looking in a flat mirror

A flat mirror is also known as a <u>plane</u> mirror. When you look into a mirror you see an **image** of your face. A mirror image is a copy of something that you see when light has had its direction changed. The light from your face has had its direction changed when it bounced off the mirror. The reflected light seems to come out of the mirror and you see an image of your face in the mirror. Because of the way light reflects off a flat mirror the image of your face follows certain rules.

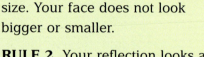
you   mirror   your image

4   How does the size of the image compare with the size of the object when you look at something in a plane mirror?

5   What happens to writing when you look at it in a mirror? Find an example of where this effect is used.

6   What are the important differences between looking at a photograph of your face and looking at yourself in a mirror?

RULE 1 The image is the same size. Your face does not look bigger or smaller.

RULE 2 Your reflection looks as if it is as far into the mirror as your face is in front of the mirror.

RULE 3 We call the thing that the light comes from the **object**. In the diagram shown here the real face is the object. The image is the opposite way around to the object.

Using two mirrors to see over the top of things

If you want to see over the top of something, you can use two mirrors in an instrument called a <u>periscope</u>. It is made from two mirrors in a tube. The light enters the periscope, reflects off the first mirror at the top, and travels down to the bottom mirror. Then the light is reflected into your eye. Periscopes are used for seeing over crowds, observing wildlife from behind walls and in submarines for looking out over the surface of the sea.

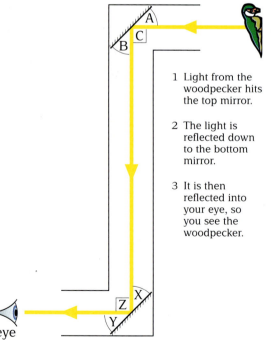

1 Light from the woodpecker hits the top mirror.

2 The light is reflected down to the bottom mirror.

3 It is then reflected into your eye, so you see the woodpecker.

eye

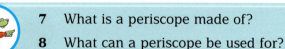

7   What is a periscope made of?

8   What can a periscope be used for?

9   What are the angles between the light rays and the mirrors in a periscope?

10   Remote control devices for televisions don't use light. They use infrared radiation. Find out how a remote control works and in what ways infrared radiation is similar to light.

8K.4 Bending light

You can change the direction of a ray of light by bouncing it off a mirror. You can also change its direction by shining it into a different transparent substance. When you shine a ray of light from one substance into another it bends. These diagrams show what happens when you shine light into a thick block of glass or a bowl of water at different angles. The dotted line drawn on the diagrams is the normal.

When light bends like this the effect is called **refraction** (refraction means bending). We say that the light has been refracted. When there is a large angle between the light ray and the normal, the refraction (the bending) is quite large. When a light ray travels along the normal there is no refraction.

1 What does the word refraction mean?
2 How does the light have to enter the glass to produce a large amount of bending?
3 How does the light have to enter the glass for no refraction to happen?

Bending the light away from the normal

Refraction works in both directions. If you shine a ray of light out of a transparent substance like water or glass into air, it bends away from the normal. The only exception to this is if you shine it along the normal, when it doesn't bend at all.

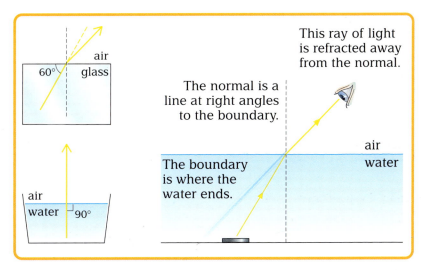

The normal is a line at right angles to the boundary.

The boundary is where the water ends.

This ray of light is refracted away from the normal.

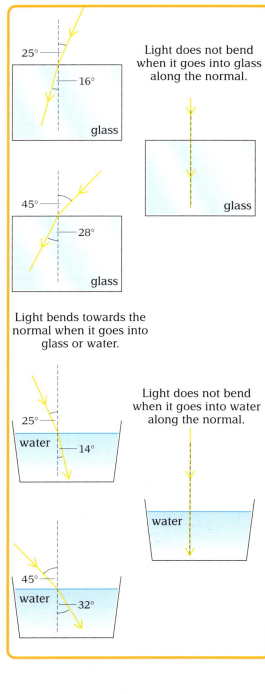

Light does not bend when it goes into glass along the normal.

Light bends towards the normal when it goes into glass or water.

Light does not bend when it goes into water along the normal.

You can use this effect to see round corners. In the top diagram Kris cannot see the coin because the light ray that travels past the edge of the metal can does not enter his eye. In the second diagram his friend Sam has added some water while Kris keeps his head still. The coin comes into view because the light ray from it is refracted (bent) as it comes out of the water. Kris can see the coin now because the light from it enters his eye.

4   How can you bend light away from the normal?

5   How must light travel out of a substance if it is not going to be refracted?

6   Explain why Kris can see the coin when the water has been added to the metal can.

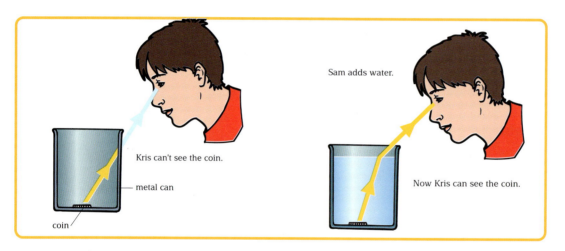

Kris can't see the coin.

metal can

coin

Sam adds water.

Now Kris can see the coin.

Water looks shallower than it is

When you look into water, refraction makes it look shallower than it is. This diagram shows what happens. The image of the fish does not appear to be as deep as the fish actually is.

This effect can be dangerous for people who cannot swim and who do not know that water seems shallower than it really is. A 2 m deep pool will only appear to be 1.5 m deep because of the refraction effect.

We can work out how deep an object is in water by multiplying how deep it appears to be by $\frac{4}{3}$.

7   Why does water appear to be shallower than it really is?

The boy looks shorter in water because light is reflected as it goes from water to air.

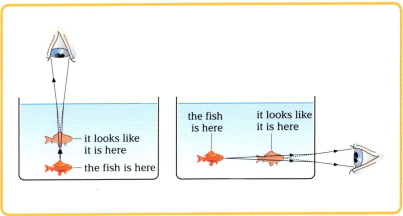

it looks like it is here
the fish is here

the fish is here
it looks like it is here

If you want to catch a goldfish by using a net, you will have to go deeper into the tank than you think if you look through the top surface. If you look through the side of the tank, the fish will appear at the correct depth but it will seem nearer! The refraction is always there to fool you when you look into water!

8 If a goldfish appears to be 15 cm deep in a fish tank, how deep is it really?

9 Find out what the critical angle is and what happens to light at the critical angle.

8K.5 The spectrum

If you go to Woolsthorpe Manor in Lincolnshire you can see a small room with a piece of glass set up to catch the Sun's rays coming through a hole in a window shutter. On the wall behind the glass the light forms a rainbow of colours. This is the room where Isaac Newton first did this experiment to show that white light is made up of colours, on 21 August 1665.

An easy way to show this is to shine white light through a specially shaped piece of glass or transparent plastic called a **prism**. The shape of the prism refracts the light twice in the same direction. The diagram shows you what happens.

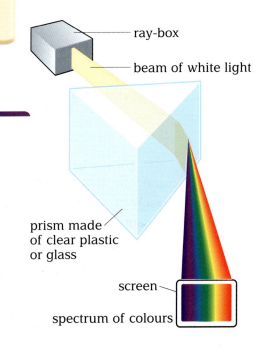

ray-box

beam of white light

prism made of clear plastic or glass

screen

spectrum of colours

The rainbow of colours is called a **spectrum**. If you look at a spectrum you will see that the colours gradually blend from one to the other, from a deep red at one end to a deep violet at the other. Many people remember the order of the colours using the phrase 'Richard Of York Gave Battle In Vain'. Each capital letter stands for a colour.

Some people say that the colour indigo is not present in the spectrum and that Newton invented it so he had seven colours rather than six. People thought that seven was a mystical number. It is a nice story but it does not make any difference. What actually matters is what is there to be seen. The seven-colour phrase is just a way of remembering the order. There are not actually any separate bands of colour in a spectrum, just a gradual change of shade.

1   What colours make up white light?
2   What is the name of the glass or plastic shape used to produce a spectrum?

3   Make up another phrase for remembering the order of colours in the spectrum.

Rainbows

Sometimes we see rainbows when the Sun is shining and it is raining at the same time. Obviously, showers of rain are not made of small prisms, but the raindrops work in a similar way.

If you are going to see a rainbow the angles have to be correct between the Sun, the rain and your eyes. You need to stand with your back to the sunlight. The angle between the direction of the sunlight and your line of sight to the raindrops must be about 42°. Red appears at the top of a rainbow and violet appears at the bottom.

4   Where does the Sun have to be for you to see a rainbow?
5   What is the order of colours in a rainbow, from the outside to the inside?

Drops of rain can split up sunlight into all the colours of the rainbow.

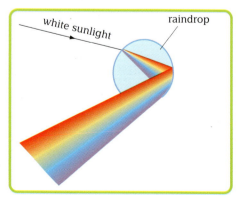

white sunlight   raindrop

You can make an artificial rainbow with a mist of drops from a garden hose.

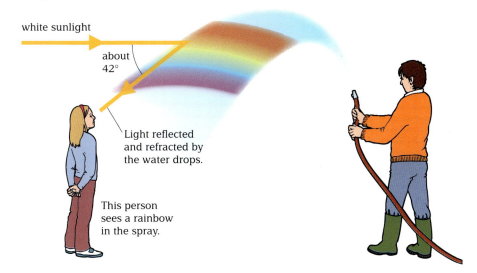

white sunlight

about 42°

Light reflected and refracted by the water drops.

This person sees a rainbow in the spray.

Another way of splitting white light into colours is to look at the reflection on the shiny side of a CD and tilt the CD at the same time. You can see some interesting rainbow patterns.

6   Give two different places where you might see a spectrum of colours that has not been produced by a prism.

7   Find out what colour blindness is.

8K.6 Colours

Coloured light can be made by passing white light through filters that only let some of the spectrum through. Coloured objects appear coloured because they only reflect certain parts of the white light spectrum.

Coloured filters

Coloured filters are sheets of plastic used to get coloured light from white light. They work by letting some of the spectrum through and absorbing other parts of it.

1   What colours does a red filter let through?

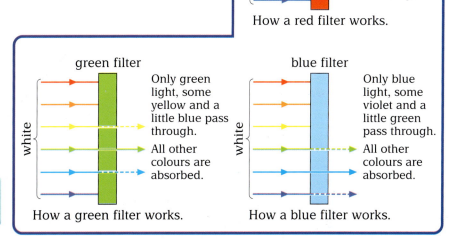

red filter

white

Only red light and a little orange pass through the filter.

The filter absorbs all other colours in the light.

How a red filter works.

green filter

white

Only green light, some yellow and a little blue pass through.

All other colours are absorbed.

How a green filter works.

blue filter

white

Only blue light, some violet and a little green pass through.

All other colours are absorbed.

How a blue filter works.

Red, green and blue are called the primary colours. There is an interesting effect that you can produce by using primary coloured filters. Because one third of the spectrum comes through the red filter, the middle third comes through the green filter and the final third comes through the blue filter, where they cross over you get white. You also get other colours where just two of the lights cross over.

This idea is used in a colour television set. It generates three pictures on the screen: one is in red dots: one is in green dots and one is in blue dots. Because the dots are very close together, when you look at the screen you see the full range of colours depending on how bright the different dots are.

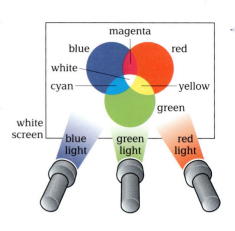

2 How many different colours of picture does a colour television have to produce for you to see the full range of colours?

3 Why does mixing red, green and blue lights produce white light?

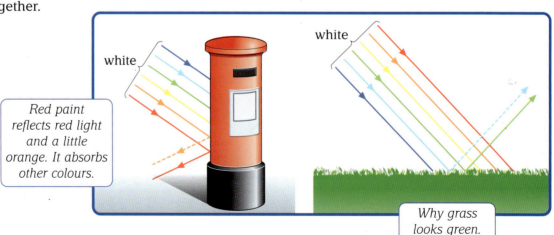

Television tube. Beams from three guns reach the screen. Each gun produces one of the three primary colours on the screen. mesh behind the screen. screen. Dots (pixels) of the primary colours form on the screen.

Why things look coloured

White light is made up of a range of colours mixed together.

Red paint reflects red light and a little orange. It absorbs other colours.

Why grass looks green.

4 What colours from the spectrum does a red postbox absorb?

5 What colours from the spectrum does grass reflect?

6 Which parts of the spectrum do you think a yellow daffodil absorbs?

As well as getting different coloured objects by reflecting parts of the spectrum, you can get objects that are shades of grey, somewhere between white and black. These diagrams show how you get white, black and grey objects.

Things look different in different light

A postbox is red because it reflects red light and absorbs other colours. If you only have a green light and you shine it on a red postbox, the postbox will look black because there is no red light for it to reflect. Green grass will look green in the green light, but it will appear black if you shine red light on it because it absorbs red.

This effect can be a problem when you are buying coloured clothes or wallpaper or paint. The lighting in a shop is usually from a fluorescent tube and the balance of colours is different from the colours in sunlight. The bar charts show the difference between some common types of bulb and the Sun.

 7 Why might a red shirt look a different shade of red when you take it outside a shop?

 8 What colour would you expect blue jeans to look under the orange light of a street lamp? Give a reason for your answer.

 9 Find out what a daylight filter is and why it is sometimes used on a camera.

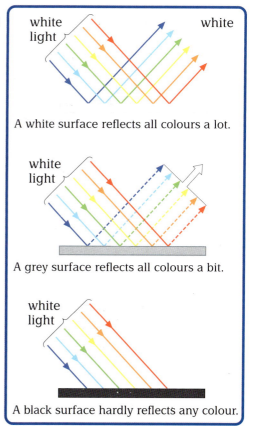

A white surface reflects all colours a lot.

A grey surface reflects all colours a bit.

A black surface hardly reflects any colour.

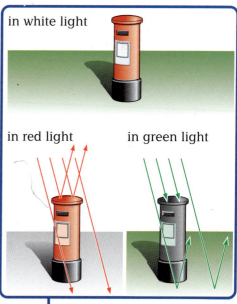

in white light

in red light   in green light

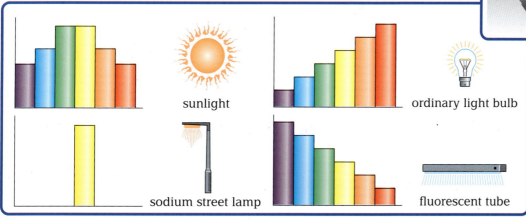

sunlight

ordinary light bulb

sodium street lamp

fluorescent tube

You should now understand the key words and key ideas shown below.

Red filters let through red light and absorb the rest of the spectrum. Other colour filters work in a similar way.

A red **object** reflects red light and absorbs other colours so it will look black in pure green light. A blue object will appear black in pure red light because it absorbs red light and reflects blue light.

Light is represented by **rays**, which are straight lines with an arrow to show the direction the light travels in.

White light can be split into a **spectrum** by a **prism**. The colours are red, orange, yellow, green, blue, indigo and violet.

Light travels in straight lines at very high speed.

Light comes from **sources** like the Sun, flames, light bulbs and hot objects. These objects are **luminous**.

Light bends when it passes from one transparent substance to another. This is called **refraction**.

Light **reflects** off some objects so we see them. These object are **non-luminous**

Light goes through **transparent** things so you can see things through them. When light goes through we say it is **transmitted**.

Light

Light is **absorbed** by objects which appear black.

Light goes through **translucent** substances but breaks up so you cannot see things clearly through them.

Light reflects off a plane mirror so that the angle of incidence equals the angle of reflection. Angles are measured from the **normal**.

The **image** in a flat mirror appears as far behind the mirror as the **object** is in front of it.

Light is stopped by **opaque** substances which cast shadows.

The image in a flat mirror is the same size as the object.

The image in a flat mirror is the same way up as the object.

Sound and hearing

In this unit we shall learn about different types of sound and how they are made. We shall also learn about how we hear sound and how loud sounds can damage our hearing.

8L.1 Making sound

Sound energy is a form of energy. It is made by things when vibrations are transformed into sound energy.

A **vibration** is a fast, backwards and forwards movement that repeats many times. Objects always vibrate to either side of their normal position. Try putting your fingers against your throat as you speak. You can feel your voice box vibrating. The kinetic energy of the vibrations is being changed into sound energy.

A loudspeaker makes sound when its paper cone vibrates backwards and forwards. This makes the air particles next to it vibrate. These vibrations are passed onto other air particles, rather like a Mexican wave. When the vibrations reach our ears, we hear the sound.

Vibrations can travel through different materials. Whales communicate by sending sound vibrations through the water. Earthquakes make vibrations in the ground that can be detected on the other side of the Earth.

You can hear the washing machine because it is vibrating.

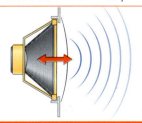

The cone of a loudspeaker vibrates backwards and forwards.

1 Look at the <u>two</u> diagrams of musical instruments. For each instrument, write down what is vibrating.

2 Name <u>three</u> materials that vibrations can travel through.

3 Explain how the sound of a drum reaches your ear. Use the word <u>vibrations</u> in your answer.

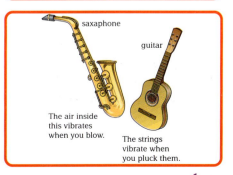

saxophone

guitar

The air inside this vibrates when you blow.

The strings vibrate when you pluck them.

Making different sounds

Sounds can vary in **loudness**. Thunder is often a very loud sound but a whisper is a very quiet or soft sound. A loud sound has more energy than a quiet sound.

Yasmin can make a loud sound on the drum by hitting it very hard. This makes the drum skin move a long way from side to side and makes big vibrations of the air particles. We say that the vibrations have large **amplitude**. If Yasmin makes a quiet sound on the drum, the vibrations of the air particles are much smaller. The sound is quiet because the vibrations have small amplitude.

4 Write down an example of a loud sound.

5 How can Yasmin make a quiet sound on the drum?

6 Choose a different musical instrument. Describe how to make a loud sound and a quiet sound using this instrument.

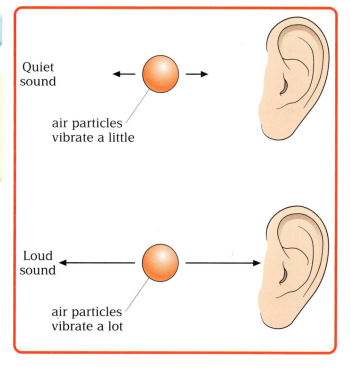

Quiet sound
air particles vibrate a little

Loud sound
air particles vibrate a lot

Sounds can also vary in **pitch**. A squeaky door produces a high-pitched sound and a big drum makes a low-pitched sound. Sometimes we can link the pitch of the sound to the size of the object that makes it. Big objects usually make low-pitched sounds because they vibrate slowly. Small objects make high-pitched sounds because they vibrate quickly.

7 Name <u>one</u> low-pitched sound and <u>one</u> high-pitched sound.

8 Look at the picture. Which instrument makes sounds of a lower pitch, the violin or the cello? Give a reason for your answer.

9 Why do you think a mouse makes a higher-pitched sound than a lion?

10 A male frog makes a low-pitched croaking sound when it is trying to attract a female frog in the spring. Find out how it makes such a low sound for its size.

Stuart's guitar has six strings of different thickness and 22 frets along its neck. If he plucks the thickest string, it produces a low-pitched sound. The thick string is heavy and so it vibrates slowly. We say that the string is vibrating at low **frequency**. The frequency is the number of vibrations in one second. It is measured in units called **hertz** (symbol Hz). The lowest note on the guitar only makes about 82 vibrations in one second, so its frequency is 82 Hz.

To make a high-pitched note he can pluck a thinner string. This vibrates faster because it is lighter. It makes a lot of vibrations in one second and we call this vibration a high frequency. The thinnest string makes about 330 vibrations in one second, so its frequency is 330 Hz.

Stuart can change the pitch and frequency of a sound in two other ways. He can press the string down against one of the frets. This makes the vibrating section of the string shorter. The shorter string vibrates at a higher frequency and makes a sound of higher pitch.

In Unit 7K we saw that the top E string on a guitar needs a force of about 75 N to make a note of the right pitch. Stuart turns the key to increase the force stretching the wire. This makes the string tighter so that the vibrations pass along its particles more quickly. Now the string vibrates at a higher frequency and produces a note of higher pitch.

11 What kind of vibration makes a high-pitched sound?

12 Describe <u>three</u> ways of making a higher pitched note on a guitar.

13 What is the frequency of the thinnest string on a guitar?

14 An oboe player plays the note A to help the orchestra tune up. This note has a frequency of 440 Hz. How many vibrations are there in one second?

Seeing sound waves

We cannot really see sound waves, but we can make a picture that represents them, using a microphone and an **oscilloscope**. The microphone changes the sound energy into tiny electric voltages which make the line on the oscilloscope move up and down.

The picture of a loud sound shows a tall wave. This is because the vibrations have large amplitude.

The picture of a quiet sound is smaller as the vibrations have small amplitude. The amplitude is shown on the oscilloscope as the height of the wave from the middle line to the crest, or top, of the wave.

The picture of a high-pitched sound shows the waves close together. This is because there are lots of vibrations in one second and the sound is of high frequency. The picture of a low-pitched sound shows the waves spread out. This is because there are fewer vibrations in one second and the sound is of low frequency.

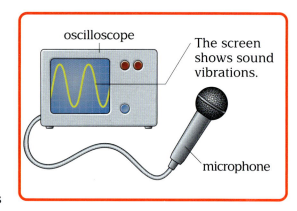

The screen shows sound vibrations.

oscilloscope

microphone

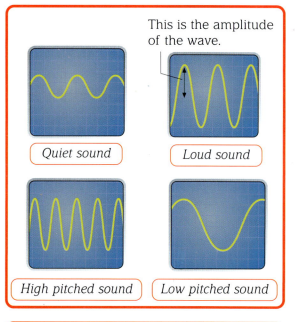

This is the amplitude of the wave.

Quiet sound Loud sound

High pitched sound Low pitched sound

15 Does a loud sound or a quiet sound produce the taller wave?

16 Which instrument, the flute or the piccolo, would produce the sound shown by:

a oscilloscope A;   **b** oscilloscope B?

17 The sound of Yasmin's voice looks like this on the oscilloscope. Explain why the wave is a complicated shape.

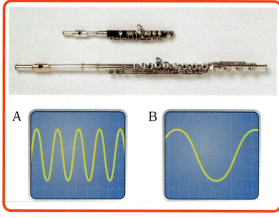

A   B

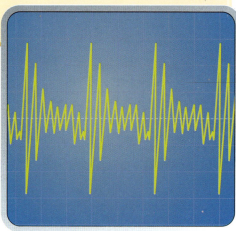

8L.2 Travelling sound

Sound travels as a series of vibrations through a material. Look at the diagram of a loudspeaker. The cone of the loudspeaker vibrates backwards and forwards. This pushes some air particles backwards and forwards. These push the next air particles, and so on, to make a sound wave.

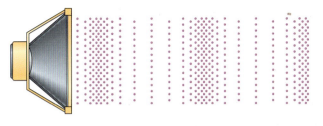

We can see this happening if we put a candle flame in front of the loudspeaker. The flame vibrates backwards and forwards at the same frequency as the loudspeaker cone.

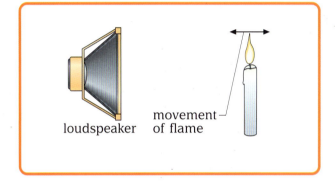

loudspeaker   movement of flame

Sound in a vacuum

In 1705, Francis Hauksbee showed a famous experiment, first performed by Robert Boyle in 1660, to an amazed audience in London. He put a bell inside a jar and made it ring. He used a pump to remove all the air from the jar to make a **vacuum**. A vacuum is an empty space that contains no particles at all. The audience could see the bell ringing but they could not hear it! Sound cannot travel through a vacuum because there are no particles to vibrate and carry the sound wave.

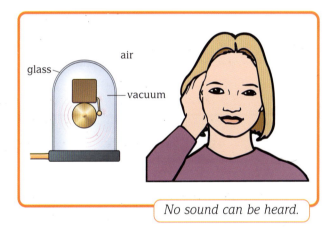

glass   air   vacuum

No sound can be heard.

1   What is a vacuum?

2   Hauksbee's audience could still see the bell vibrating. What does this tell us about light?

3   The Sun is an exploding ball of hot gases with a surface temperature of 5500 °C. Space is a vacuum. Why can we see the Sun but not hear the sound it makes?

4   Astronauts sometimes leave their spacecraft to mend a faulty satellite. Why do they use a radio system to talk to each other?

The speed of sound

In air, sound travels at a speed of about 330 metres every second. That is very fast – over 30 times faster than the world's fastest athlete!

The table shows that sound travels faster in water than in air. It travels even faster in solids like brick and iron. In Unit 7G we learnt that the particles in solids and liquids are closer together than those in a gas. The vibrations pass from particle to particle more quickly, so the wave travels faster.

Material	Speed of sound in m/s
air	330
water	1500
brick	3000
iron	5000

5 Draw a bar chart to show the speed of sound in different materials.

6 When a train is approaching a station we can usually hear the railway lines making a humming sound. Explain why we can hear this before we hear the train itself.

7 A whale sends a message to her mate 150 000 metres away. How long will the message take to reach him?

During a thunderstorm we see the lightning much earlier than we hear the thunder. Sound travels fast but light travels nearly a million times faster, at an amazing 300 million metres every second! Light could travel more than seven times round the Earth in one second!

8 How much faster does light travel than sound?

9 Describe <u>two</u> more examples which show that light travels much faster than sound.

10 If you marked the speed of light on your bar chart from question 5, how long would the bar be?

11 Describe how you could measure the speed of sound in air and in water. How would you make sure your value was accurate?

8L.3 Hearing sound

Look at the pictures of animals. They all have two ears but they can hear very different sounds.

A blackbird can hear very quiet sounds. It runs up and down on the grass to make worms think that it is raining. Then it listens for the sound of a worm moving in the ground so that it knows where to dig.

A dog can hear very high-pitched sounds. Dog owners sometimes blow on a special high-pitched whistle to call their dog even though they might not be able to hear the sound themselves.

A bat hears even higher sounds made by the insects that it eats. It can make very high-pitched sounds to find its way around in the dark. It listens for the echoes that are made when the sounds reflect off walls and other objects. This is called <u>echolocation</u>.

The bar chart shows the range of frequencies that different animals can hear. People can only hear sounds between 20 Hz and 20 000 Hz. This is a very small range compared with most animals. This range usually gets smaller as people get older. Your teacher will probably not hear the high-pitched notes as well as you do. Some people can hear fewer sounds, either because their hearing has been damaged by loud sounds or because of other causes.

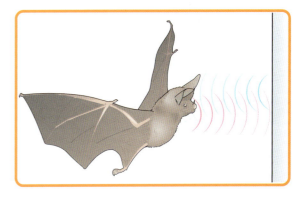

1. Write a list of animals in order of the highest frequency they can hear.

2. As people grow older, which types of sound do they find it difficult to hear?

3. Make a list of sounds that an older person might not be able to hear. Test them out if you can.

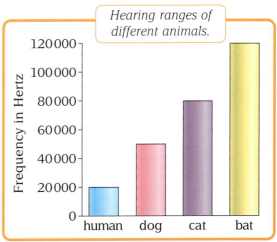

Hearing ranges of different animals.

The ear

Your ear is a very sensitive part of your body.

- A sound wave travels down the canal and makes the eardrum vibrate.
- These vibrations are passed onto the cochlea by a set of three small bones.
- In the cochlea, a liquid moves backwards and forwards and stimulates the nerve cells inside it.
- The nerve cells make small electrical signals.
- These electrical signals travel along the nerve to the brain.
- The brain receives an exact copy of the vibrations that made the sound wave.

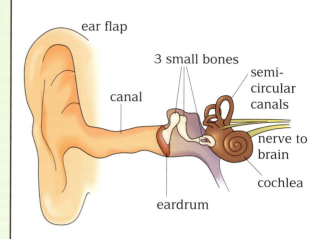

ear flap

3 small bones

semi-circular canals

canal

nerve to brain

cochlea

eardrum

4 What are the sound vibrations changed to before they reach the brain?

5 Draw a flow diagram for the ear, to show how vibrations in a sound wave send messages to the brain.

All animals have two ears. It helps the animal to find out where the sound is coming from. The owl hears the sound of its prey in both ears but the sound waves take slightly longer to reach its right ear. The owl's brain works out how far away its prey is. Then the owl turns his head round to make another measurement. Now he can swoop down and catch his prey.

The long-eared bat has huge earflaps, which collect more sound vibrations in the air. It can hear very quiet sounds and so it only needs to make quiet sounds for echolocation. This means that its prey insects cannot hear it coming!

6 Why do animals have two ears?

7 Which type of bat can hear the quieter sounds, the pipistrelle or the long-eared bat?

8 Find out how a stethoscope helps doctors to hear what is going on inside a patient.

long-eared bat   *pipstrelle bat*

8L.4 Dangerous sound

Loud sounds are made up of large vibrations, so the sound waves can carry a lot of energy. Very loud sounds can damage the ear and sometimes they cause pain. The loudness of a sound can be measured using a sound level meter. It is measured in units called **decibels**. The symbol for this unit is dB.

1 Which loudness level can permanently damage the ear? Give your answer in dB.

2 How loud is your classroom at the moment? Use the chart to judge the value.

3 People describe some sounds as noise. What do you think they mean by noise?

Making sound quieter

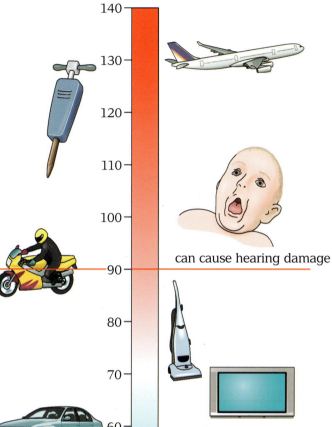

can cause hearing damage

The easiest way to make a sound quieter is to move away from the source – or to turn it down!

It is not always possible to do this, so other methods are used.

Cars have an exhaust pipe fitted to the engine and this has a silencer on it. This does not get rid of all the sound of the engine but it does make it a lot quieter. A car makes a very loud sound if it has a hole in its exhaust pipe!

At home, a room sounds very different when the curtains and carpets are taken out. This is because soft materials absorb sound vibrations and stop the sound from reflecting to make an echo.

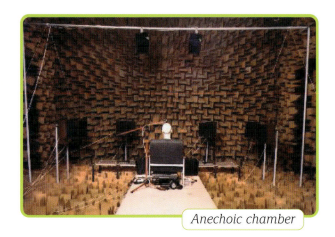

Anechoic chamber

An anechoic chamber is designed to absorb all sound. It is used for making very accurate measurements and recordings of sound.

The Royal Albert Hall in London has some strange mushroom-shaped saucers that hang from the ceiling to stop echoes. Recording studios and concert halls often have double glazing, or even triple glazing, installed to stop traffic noise from outside disturbing the performances.

Symphony Hall in Birmingham was built in 1991 and experts say that it has very good acoustics. This means that there is not very much unwanted sound and it is easy to hear the performers clearly.

There are strict laws that limit the noise level near places where people live and work. Too much noise is called **noise pollution**.

People working near loud machines wear earplugs or ear defenders. These absorb some of the sound vibrations.

Concorde can fly three times faster than sound. It is not allowed to fly faster than the speed of sound while it is over land. This is because it makes a very loud sound called a sonic boom.

4   What is meant by noise pollution?

5   Describe how to reduce echoes and traffic noise in a concert hall.

6   Why do people sometimes wear ear defenders at work?

7   Soft materials absorb sound vibrations. What happens to the energy of the sound waves?

Damaged hearing

Evelyn Glennie is a famous percussion player. She plays with different orchestras in concert halls all over the world. This is amazing as she became profoundly deaf when she was 12 years old. She says that she can feel the vibrations of the music through her body. In conversation, she speaks clearly and works out what other people are saying by watching their lips move. This is called <u>lip reading</u>.

People with very poor hearing are said to be deaf. You might be surprised to know that most deaf people can still hear some sounds.

There are many reasons why people have hearing problems. Some people are born with poor hearing. Others find that their hearing becomes worse after an illness or ear infection.

If the canal is blocked by earwax the vibrations cannot reach the eardrum. It is dangerous to clean out the earwax using a stick or cotton bud as this might damage the ear drum. The ear drum is a very thin layer of skin and muscle and so it breaks easily.

If the nerve cells in the cochlea are damaged, they do not send clear electrical signals to the brain. This is the most common cause of hearing problems in older people. Loud sounds can also damage the nerve cells in the cochlea. If you go to a loud pop concert or disco, you might become deaf for a few days. Sometimes the loud sounds cause a very annoying ringing sound in the ears called tinnitus. Both of these effects should go away after a few days. If you listen to very loud sounds for a long time, it can damage your hearing permanently.

8   What most frequently causes older people to have hearing problems?

9   Describe <u>one</u> way of breaking the ear drum.

10 Evelyn Glennie is profoundly deaf. Find out what she can hear.

11 Find out more about the life of Evelyn Glennie by searching on the Internet.

12 Deaf people often use a sign language with their hands. Find out how to say a simple phrase using this sign language.

You should now understand the key words and key ideas shown below.

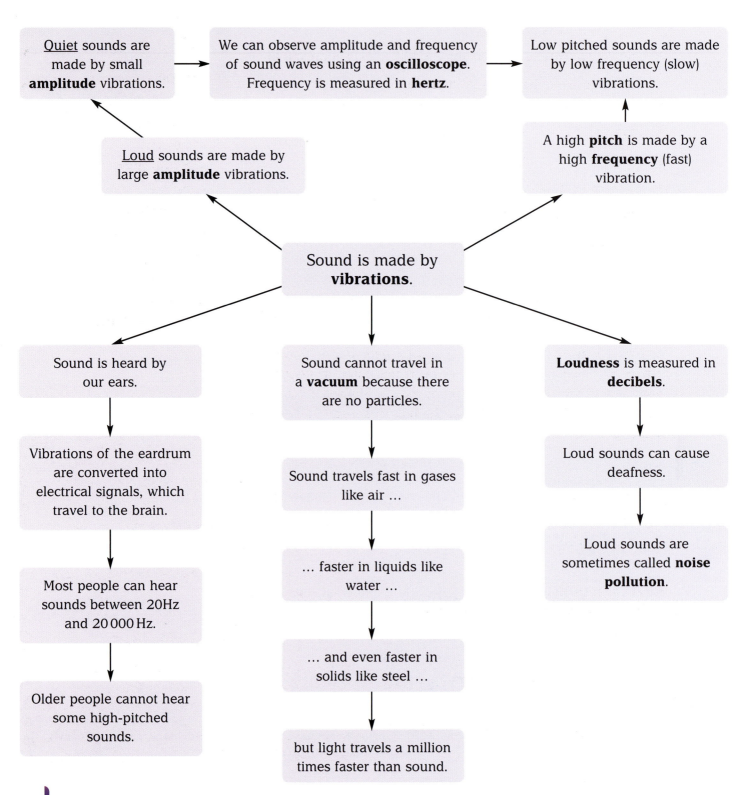

Quiet sounds are made by small **amplitude** vibrations.

We can observe amplitude and frequency of sound waves using an **oscilloscope**. Frequency is measured in **hertz**.

Low pitched sounds are made by low frequency (slow) vibrations.

Loud sounds are made by large **amplitude** vibrations.

A high **pitch** is made by a high **frequency** (fast) vibration.

Sound is made by **vibrations**.

Sound is heard by our ears.

Sound cannot travel in a **vacuum** because there are no particles.

Loudness is measured in **decibels**.

Vibrations of the eardrum are converted into electrical signals, which travel to the brain.

Sound travels fast in gases like air …

Loud sounds can cause deafness.

Most people can hear sounds between 20 Hz and 20 000 Hz.

… faster in liquids like water …

Loud sounds are sometimes called **noise pollution**.

Older people cannot hear some high-pitched sounds.

… and even faster in solids like steel …

but light travels a million times faster than sound.

Scientific investigations

In Year 7 you learnt and used some basic investigation skills.

These pages are to help you to improve these skills and to introduce some new skills. We are going to look at:

- choosing the best strategy for an investigation;
- using computers to collect data and present results;
- choosing a suitable range of readings;
- using scientific ideas in conclusions;
- writing a more detailed evaluation.

Choosing the best strategy for an investigation

Often there are several possible ways of getting results when you carry out an investigation. This means there is often no right way or wrong way. However, one method may be better than the others. For example, one way may be easier, safer or produce more useful data than all the other methods. Molly and Harry were asked to find out which was the best of several materials to insulate the loft of a house:

KEY WORDS

reliable
secondary data
sampling
surveys
sample size
random
data collection
presenting results
range
precision
evidence
anomalous results

- Use a beaker of water that has just boiled.
- Wrap the insulating material around the beaker.
- Place a thermometer inside the beaker and measure the temperature of the water every 2 minutes for 20 minutes.
- Repeat this for the different materials.

Molly's ideas

- Use an upside down shoebox as a model of the house.
- Heat the house using a beaker of water that has just boiled.
- Place the material to be tested on top of the shoebox.
- Put an ice cream tub on top of the material to represent the roof.
- Use a thermometer to measure the temperature inside the roof every 2 minutes for 20 minutes.
- Repeat this for the different materials.

Harry's ideas

1   Draw pictures of Molly's apparatus and Harry's apparatus.

2   What must they do to make sure the test is fair for the different materials?'

3   Which method do you think will give the best results? Explain why.

Sharing data

Often you can use a range of sources of information and data in an investigation. The more data you have for an investigation, the easier it is to be confident about your conclusion.

The results of Molly's and Harry's investigation will be more **reliable** if they both agree on the same experiment and share their data. A whole class set of results produces even more accurate and reliable data. So Molly and Harry can be even more confident about their conclusions.

4   Why is it more accurate to have several sets of data?

5   What else do you need to consider when using someone else's data?

Secondary sources, sampling and surveys

A laboratory investigation is not the only form of scientific enquiry that you can do.

You can:

- use secondary sources. If you use information collected by other people to help you to answer a question, it is called **secondary data**. For example you can get information about loft insulation materials from leaflets produced by energy companies, "Which" reports, libraries and the Internet. However, some of this information may be biased and suggest that their results are better than they really are.

- use sampling to do surveys. There are examples of plant and animal **sampling** in unit 8D. **Surveys** of data about people, including people's opinions, often involve questionnaires.

6   Which type of secondary source of information is the most likely to be biased? Explain your answer.

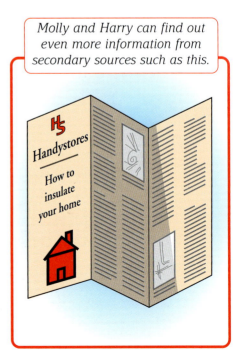

Molly and Harry can find out even more information from secondary sources such as this.

Handystores

How to insulate your home

7 What is secondary data?

8 List some secondary sources of information about factors affecting the pH of soil.

In Year 7, you learnt about the importance of controlling variables in investigations. But some variables are easier to control than others. Physical factors such as light and temperature are fairly easy to control. Living things are a particular problem because they themselves vary. So when you investigate living things, you use 10 or 20 or more, not just use one. We call this using a sample. You have to think about a suitable **sample size**.

9 Why do you use a sample when you investigate living things?

In Unit 8D you learn how to estimate the numbers of different living things by sampling using quadrats. Factors such as light and amount of water affect what grows.

10 Why is it better to use 20 quadrats than five?

11 Explain why the sample needs to be random.

12 Describe the <u>primary</u> sources that Martin will use when he writes up his report about the proposed airport.

*By taking a sample of 20 **random** quadrats, these pupils are allowing for the fact that conditions are not the same in all parts of the beach.*

Martin is a scientist using scientific books and papers to do a desk study about the environmental impact of a planned new airport.

Martin uses traps to sample animals, and quadrats to sample plants, on the land where the airport is planned.

Using computers to collect data and present results

Using the Internet is just one example of using computers to help with investigations. We are now going to look at two more: **data collection** and **presenting results**.

Gill and Gita were studying their school pond. They looked at all of the different species in the pond and drew a food web for these. Then Gita suggested that, as the amount of light falling on the pond was lower at night, the pond water would contain less oxygen at night.

They decided to investigate the changes over a period of 24 hours in:

- the amount of dissolved oxygen;
- the temperature of the water;
- the amount of light falling on the water.

Gill suggested that the best way to take these readings was to use a datalogger. Then they could collect data over 24 hours without having to remember to take readings or having to stay up all night.

 13 What are the advantages in using a datalogger for this experiment?

You can plot the results of experiments, whether collected by hand or with a datalogger, using a spreadsheet program. Careful plotting can make results easier to read. When you get used to using a spreadsheet for plotting graphs, it will also save you time.

 14 Write down <u>two</u> advantages of plotting a graph using a spreadsheet package.

 15 Give <u>one</u> disadvantage of plotting a graph using a spreadsheet package.

Choosing a suitable range of readings

Daniel and Lauren want to investigate how the strength of an acid affects the time it takes for a lump of rock to dissolve.

First, they need to choose suitable concentrations of acid to use.

100 cm³ Acid and 0 cm³ water
90 cm³ Acid and 10 cm³ water
80 cm³ Acid and 20 cm³ water
70 cm³ Acid and 30 cm³ water
60 cm³ Acid and 40 cm³ water
50 cm³ Acid and 50 cm³ water
40 cm³ Acid and 60 cm³ water
30 cm³ Acid and 70 cm³ water
20 cm³ Acid and 80 cm³ water
10 cm³ Acid and 90 cm³ water
0 cm³ Acid and 100 cm³ water

These are the acid concentrations Lauren suggested.

80 cm³ Acid and 20 cm³ water
81 cm³ Acid and 19 cm³ water
82 cm³ Acid and 18 cm³ water
83 cm³ Acid and 17 cm³ water
84 cm³ Acid and 16 cm³ water
85 cm³ Acid and 15 cm³ water
86 cm³ Acid and 14 cm³ water
87 cm³ Acid and 13 cm³ water
88 cm³ Acid and 12 cm³ water
89 cm³ Acid and 11 cm³ water
90 cm³ Acid and 10 cm³ water

These are the acid concentrations Daniel suggested.

Lauren suggested that it would be hard to spot a pattern in Daniel's readings because his acid concentrations do not cover a wide enough **range**.

16 What are the advantages and disadvantages of Lauren's range of acid concentrations?

It is also important to consider the **precision** of your results. For example, you wouldn't be able to use kitchen scales to measure the mass of a small strip of magnesium, you would not get a useful reading. A chemical balance would be much more useful.

17 What apparatus can Daniel and Lauren use to measure precise volumes of acid and water?

Using scientific ideas in conclusions

In Year 7 you started to look at how to draw conclusions from your results. A scientist then tries to explain the results using scientific ideas.

> *Harry wrote this conclusion for his insulation experiment.*

In my experiment, the temperature in the loft space was the lowest when I used fibre-glass as my insulator. This means fibre-glass is the best insulator. I think this is because it traps heat in-between the fibres.

> *Molly wrote this conclusion.*

In my experiment, the temperature of the water in the beaker stayed highest when I used fibre-glass as the insulator. Fibreglass is the best insulator. This is because the fibreglass traps air molecules in its fibres. Air is not a good conductor of heat energy. So it is this air that insulates the beaker.

 18 Whose conclusion uses the more detailed scientific ideas?

Writing a more detailed evaluation

In Year 7 you started to write evaluations of your investigations. You now need to try to make these more detailed. You need to consider how you could improve your investigation to increase the strength of your **evidence**.

For example, perhaps you could have:

- taken readings more accurately;
- taken more readings;
- repeated readings.

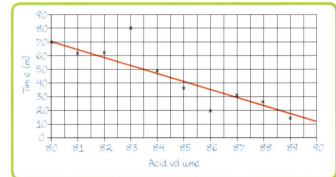

You also need to try to spot any results that do not fit the general pattern. These are called **anomalous results**.

Look at the graph showing a line of best fit.

 19 Which points on the graph do not lie near the line of best fit?

 20 What can you do to improve this set of readings?

Glossary/Index

Words in *italics* are themselves defined in the glossary.

A

abrasion   wear caused by one substance rubbing against another   75, 82, 86

absorb, absorption   when an object or substance takes in another object or substance. When living *cells* or *blood* take in *dissolved nutrients* or *oxygen* 1, 7-8, 11-12, 26, 41, 70, 123-124

acid   a *solution* that reacts with many metals to produce a *salt* and *hydrogen*, and with *alkalis* to produce a *salt* and water   10, 30, 34, 66, 70, 77, 89

acoustics   how a building reflects the sounds produced inside it   144

adapt, adaptation   when plants or animals have characteristics which make them suitable for where they live   39, 50

aerobic respiration   using *oxygen* to break down food to release *energy*   13-14, 19, 21-22, 26

air   a *mixture* of *gases*, mainly *nitrogen* and *oxygen* 15, 17, 19-20, 22-26, 54, 63-4, 69-72, 74, 77

air sacs   small sacs made of groups of *alveoli* at the end of air tubes in the *lungs*   20

alveoli   tiny round parts that make up the *air sacs* in the *lungs*; one is an alveolus   20-21

amino acids   carbon compounds that *proteins* are built from   9, 11, 13, 26

amplitude   the distance from the centre position of a wave to the peak; the loudness of sound 136, 138, 146

anechoic chamber   a room which will *absorb* all sound, with no echos   144

angle of incidence   the angle between an *incident ray* and the *normal*   125

angle of reflection   the angle between a *reflected ray* and the *normal*   125

Anning, Mary   1799-1847   85

antibiotic   *drugs* used to kill *bacteria* in the body 27, 30, 35-36, 38

antibodies   chemicals made by *white blood cells* to destroy *bacteria* and other *micro-organisms* 35, 37

anus   the opening at the end of the *digestive system*   8-9

argon   an *inert* or *noble gas* that makes up about 1% of the air   63, 69-70, 72, 74

armature   a moving *magnetic* metal part of an *electromagnetic* device   118

arteries   blood vessels that carry *blood* away from the *heart*   13, 16-18, 26

atom   the smallest *particle* of an *element*   51, 54-55, 57, 59-60, 62-64, 67, 74

attract, attraction   two objects pulling towards each other   110, 111

B

bacteria   *micro-organisms* that are *cells* without true *nuclei*; one is called a bacterium   11, 27-28, 31, 33-37

basalt   an *igneous rock* with small grains because it formed from a *magma* that cooled quickly   78, 87, 92-93, 96

base   also known as an alkali   66

blood   a *liquid* in animal circulatory systems   7-8, 11-20, 26, 35, 70

blood system   the *heart, arteries, capillaries, veins* and the *blood* that *circulates* through them   16

boiling point   the temperature at which a *liquid* changes into a *gas*   63, 71-74, 108, 109

breathe, breathing   taking *air* in and out of the *lungs*   4, 13, 15, 20, 22, 24, 26, 34, 63, 69-70

burning   when substances react with *oxygen* and release *energy*; also called *combustion*   14, 58, 66, 71

C

calcium   a metal *element*; a *mineral nutrient* that living things need   4-5

capillary   a narrow *blood* vessel with walls only one *cell* thick   13, 15-18, 20-21, 26

carbohydrates   carbon compounds used by living things as an *energy* source, for example *starch* and sugars   1, 3-7, 12

carbon dioxide   a *gas* in the *air* produced by living things in *respiration*, in *combustion* or *burning* and when an *acid reacts* with a *carbonate*   13-16, 20, 22-26, 29, 55, 57, 60-61, 63-64, 66, 69, 71, 74, 77, 89

carbonate   compounds that *react* with *acids* to produce *carbon dioxide*; *limestone* is calcium carbonate   60, 64, 66, 77, 84, 89

carnivore   an animal that eats flesh   39, 47, 50

cell (in biology)   building block of plants and animals   1-3, 9, 11-17, 19-20, 22, 24, 28, 52

cell wall   outer supporting layer of a plant *cell*   5, 28

Celsius   a *temperature* scale with the *melting point* of ice written as as 0 °C and the *boiling point* of water as 100 °C   98

Celsius, Anders   1701-1744   98

Chain, Ernst   1906-1979   36

chemical change, reaction   a *reaction* between chemicals; it produces a new substance   14, 19, 58, 61-62, 65-66 87, 91

chemical equation   a *word equation* or *symbol equation* that shows what happens in a *chemical reaction*   61-62

chemical weathering   when *chemical reactions* cause the *weathering* or breakdown of rocks   75, 77-78, 86

circulate, circulation   the flow of *blood* around the body through the *heart, arteries, capillaries* and *veins*   16-18

classification   sorting things into groups   39-40, 50, 95

cochlea   part of the inner ear; converts movement into electrical pulses which travel along nerves to the brain   142

coil   wire wound round and round; used in *electromagnets*   116, 117, 119

combustion   when substances react with *oxygen* and release *energy*; another word for *burning*   66

community   all the plants and animals that live in a particular place   39, 42, 45-46, 48, 50

compass   a device which has a magnetic needle which points towards the North Pole   112

compound   a substance made from the *atoms* of two or more different *elements* joined together   51, 59-67, 69, 73-74

concentration   the strength of a *solution* or the amount of a substance in a *mixture*   15, 84

condense, condensation   when a *gas* cools and changes into a *liquid*   71, 108, 109

conduction   in heat conduction, energy passes along a *solid* as its *particles* heat up and vibrate faster   100-102, 105, 106, 109

conductor, of heat energy   a substance that will let *heat energy* pass through it   100-102, 109

consumer   an animal that cannot make its own food, but eats plants and other animals   39, 46-47, 50

contract   become smaller; *solids*, *liquids* and *gases* do this when they cool   58, 80

control   part of an experiment that is needed to make a test fair; it is needed so that we can be sure of the cause of a change or a difference   25

convection   a method of heat transfer in fluids (*liquids* and *gases*); the fluids at a higher *temperature* rise towards the top of the container they are in   102-103, 105-106, 109

cooling curve   a graph with time on the x-axis and *temperature* on the y-axis for a substance whose *temperature* is falling   107

core   the centre of an object; an *electromagnet* usually has an iron core   116, 119

crystal   a substance that forms from a *melted* or a *dissolved solid* in a definite shape   52, 55, 59, 65, 77, 83-85, 92-93

D

Dalton, John   1766-1844   54

decibel   unit used to measure the loudness of sound   143

Democritus   about 460-370BC   54

deposit, deposition   when *eroded* rock fragments settle   75, 81, 83-84, 86, 95

diatomic   made up of two *atoms*; O_2 is an example   61

diffusion   the spreading out of a gas or a *dissolved* substance because its *particles* are moving at random   13, 15, 20, 26

digest, digestion   the breakdown of large, *insoluble molecules* into small *soluble* ones which can be *absorbed*   1, 7, 9, 11-13, 26, 58

digestive system   all the *organs* which are used to *digest* food   4-5, 7-10, 12, 14

disease   when some part of a plant or animal isn't working properly   4, 27-28, 30-2, 34-38, 48

dissolve   when the *particles* of a substance completely mix with the *particles* of a *liquid* to make a clear *solution*   68-69, 77-78, 83-84, 86

distillation   *boiling* or *evaporating* a *liquid* and then *condensing* it to get a *pure liquid*   72

drug   a substance that can change the way that your body works or to treat a *disease*   36

E

eardrum   a tightly stretch skin found at the end of the ear canal; it *vibrates* when sound reaches it   142, 145

echolocation   a method used to judge the distance to objects, used by bats. This involves sending out a high *pitched* sound which *reflects* off objects   141, 142

egest   get rid of *faeces* from the *digestive system*   11

electrodes   these are connected to a power supply and put into a *melted* or *dissolved* substance so that *electrolysis* can happen   61

electrolysis   the process of splitting up a *melted* or *dissolved compound* by passing an electric current through it   61

electromagnet   a device which becomes *magnetic* when an electric current flows through a wire *coil*   116-118, 119

element   a substance that can't be split into anything simpler by *chemical reactions*   51-53, 55-60, 62-65, 67, 73-74

energy   energy is needed to make things happen   1-15, 19, 26,29, 49, 70, 81-82

environment   the surroundings or conditions in which plants and animals live   39, 41, 50, 85

environmental conditions conditions such as light level and *temperature* in the *environment* 39, 42, 45-46, 50

enzymes *protein* substances made in cells; they speed up *chemical reactions* 1, 9-10, 12

epidemic an outbreak of a *disease* affecting a large number of people 32-33

Erasistratus about 304-250 BC 18

erosion wearing away of rocks involving the movement of the rock fragments away from where they formed 75, 81, 86, 95-96

erupt, eruption when *lava*, *volcanic ash* and *gases* come out onto the surface of the Earth 87, 93-94, 96

evaporate, evaporation when a *liquid* changes into a *gas* 69, 84, 86, 92, 108, 109

evidence observations and measurements on which theories are based 33, 54, 94, 147, 152

expand, expansion when a substance gets bigger because its *particles* speed up and move further apart 58, 72, 79, 80, 86

F

faeces undigested waste that passes out through the *anus* 1, 8, 11-12

fats part of our food that we use for *energy* 1, 3, 5-7, 9, 11-12

fatty acid one of the building blocks of *fats* 9, 11-12

fibre undigestible cellulose in our food; it prevents constipation 1, 5-6, 11-12

Fleming, Alexander 1881-1955 36

Florey, Howard 1898-1968 36

food chain a diagram showing what animals eat 39, 46-47, 49-50

food web a diagram showing what eats what in a *habitat* 39, 47-48, 50

force a push or a pull 80-81

formula uses *symbols* to show how many *atoms* of *elements* are joined together to form a *molecule* of an element or a *compound* 51, 60-62, 64, 74

fossil remains of plants and animals from long ago 75, 84-88, 96

fractional distillation the separation of a *mixture* of *liquids* by *distillation* 63, 72, 74

freezing when a *liquid* cools and becomes *solid* 58, 79, 86

freezing point the temperature at which a *liquid* becomes a *solid* 92

frequency the number of waves (for example sound waves) produced every second 137-138, 141, 146

fret part of a guitar, used to reduce the length of string vibrating 137

fuel a substance that *burns* to release energy 14

fungi a group of living things, including *micro-organisms* such as moulds and *yeasts* which cannot make their own food; one is called a fungus 27-28, 30-31, 35, 45

G

gabbro an *igneous rock* with large grains because it formed from a *magma* that cooled slowly under the ground 87, 94, 96

Galen, Claudius about 130-200 18

Galileo Galilei 1564-1642 98, 122

gas a substance that spreads out (*diffuses*) to fill all the space available, but can be compressed into a smaller volume 20, 54, 56, 59, 61, 63-64, 66, 69-74, 76, 93

gas exchange taking useful *gases* into a body or *cell* and getting rid of waste gases 13, 20-21, 26

Gilbert, William 1544-1603 114

glucose a *carbohydrate* that is a small, *soluble molecule* (a sugar)   7, 10, 13-15, 19, 26

glycerol one of the building blocks of *fats*   9, 11-12

granite an *igneous rock* with large grains because it formed from a *magma* that cooled slowly under the ground   75-78, 86-87, 92-94, 96

gravity the *attraction* of bodies towards each other   81

growth becoming bigger and more complicated   1-3, 11-14, 26, 29-30, 35, 49

H

habitat the place where a plant or animal lives   39, 41-42, 44-45, 50

Harvey, William 1578-1657   17-18

Hauksbee, Francis 1666-1713   139

heart an *organ* which pumps *blood*   4, 13, 16-17, 26, 70

heat energy *energy* possessed by hot objects   99-106, 109

herbivore an animal that eats plants   39, 47, 50

Hertz, Hz unit of *frequency*   137, 141

Hutton, James 1726-1797   85, 87

hydrochloric acid an *acid* produced by *dissolving* hydrogen chloride *gas* in water   60, 89

hydrogen a flammable *gas*; it *burns* to form water   52, 55-57, 59, 61, 64-65, 67, 73

hydrogencarbonate indicator an indicator that detects *carbon dioxide*   24

I

Ibn-al-Nafis also known as Al-Quarashi   1213-1288   18

igneous rock formed when ash or molten *magma* from inside the Earth cools and *solidifies*   75, 77-78, 86-87, 92, 94-95

image the *reflection* of an object in a mirror   126, 128, 134

immune, immunity able to resist an infectious *disease* as a result of *immunisation* or of having had the disease   27, 35, 37-38

immunisation an injection given to give the patient *immunity*   27, 35, 37-38

incident ray a beam of light which goes towards an optical device such as a mirror or *prism*   125

inert gases a group of unreactive *gases*; also called *noble gases*   63

infect, infection when *micro-organisms* get into your body and cause a *disease*   27, 31-32, 36

insoluble a word to describe a substance that will not *dissolve*   7, 66

insulation, insulator, insulate (of heat energy) material that does not *conduct* heat; it prevents heat loss   100-101, 105-106

invertebrate an animal without a backbone   39-40, 50

iron an *element*; a common metal   4-5, 12, 51, 54-58, 63-66, 73, 78, 94, 110, 116, 117, 118, 119

J

joule, J *energy* or *work* is measured in units called joules   99

K

kilojoule 1000 *joules*   4

L

large intestine the wide part of the intestine between the *small intestine* and *anus*   8, 10

lava *igneous rock* that *solidifies* on the Earth's surface   87, 91, 93-94, 96

limestone a *sedimentary rock* made from *calcium carbonate*   75, 77, 79, 84-86, 88-90, 96

lime-water a *solution* used to test for *carbon dioxide*; carbon dioxide turns the clear *solution* cloudy   22, 24-25

liquefied changed from a *solid* to a *liquid*   63, 72, 74

liquid a substance that has a fixed volume but takes the shape of its container   17, 34, 51, 54, 56, 72-73

liver large *organ* in the lower part of your body that makes bile and stores *energy*   8, 11

lodestone a naturally occurring stone which is *magnetic*   110

loudspeaker a device which converts electrical energy into sound energy   139

luminous gives out its own light   124, 134

lungs *organs* for gas exchange between the blood and the *air*   13, 15-16, 20, 26, 34, 70

M

magma molten rock from below the Earth's crust   77, 86-87, 92-96

magnet something which *attracts* a *magnetic material*   110-119

magnetic field the area around a *magnet* where it exerts a *force* on a *magnetic material*   113-114, 117, 119

magnetic force the *force* a *magnet* exerts on a *magnetic material*   110-113, 119

magnetic material something which will be *attracted* to a *magnet*   110-111, 119

magnetic tape used in devices such as a cassette recorder to record data   115

Malpighi, Marcello 1628-1694   18

mass the amount of stuff something is made of   68, 69

material substances from which objects are made   1-3, 12-13, 51-54, 58-59, 62, 75, 87

melting when a *solid* heats up and changes into a *liquid*   58, 73-74, 77, 79, 90-92, 95-96

melting point the temperature at which a *solid* turns into a *liquid*   63, 73-74, 92, 107, 109

metamorhic rocks rock formed from another type of rock as a result of heat and/or *pressure*   87, 90, 95-96

metamorphism the process of producing a *metamorphic* rock   90

microbe another word for a *micro-organism*   27

micro-organism a microscopic living thing; some cause *disease*   27-31, 34-35, 38

microphone a device which converts sound into electrical signals   138

minerals (i) simple chemicals that plants and animals need to stay healthy

(ii) the chemicals in rocks   1, 4-5, 7, 11-12, 30, 68, 74-75, 77-78, 86, 90, 92, 94-95

mixture a substance in which two or more substances are mixed but not joined together   63, 67-69, 71-75, 82, 86, 91

model in the mind, it is a group of ideas and pictures   7, 59, 62, 64

molecule the smallest part of a chemical compound and the smallest part of an *element* that can exist in nature   7, 9, 12, 19, 51, 55, 57, 60-62

muscles *tissues* and *organs* that contract to cause movement   4-5, 8, 13, 15-16

N

neutralise, neutralisation when an *acid* reacts with an *alkali* to make a neutral solution of a *salt* in water   66

Newton, Isaac 1642-1727   129-130

nitrogen a *gas* that makes up about four fifths of the *air* 23, 56-57, 61, 63, 69-70, 72, 74

noble gases another word for *inert gases*; a group of unreactive *gases* 70, 74

non-porous describes the *texture* of a rock without *pores* 75-77, 88

non-vascular plants a group of plants without a specialised transport or vascular system 40-41, 50

normal a line drawn at 90o to the point where a *ray of light* hits a mirror or a *prism* 125, 127, 134

north-seeking pole the end of a *magnet* which if left to spin freely will point towards the Earth's North Pole 112

nucleus the part of a *cell* which controls what happens in the cell 28

nutrients the food materials that *cells* use 1, 5-8, 11, 46

O

object something which if placed in front of a mirror will form an *image* on the other side of the mirror 126, 134

obsidian an *igneous rock* that cools so quickly that it is glassy, not crystalline; also called volcanic glass 87, 93, 96

oesophagus the tube between your mouth and your *stomach*; also called the gullet 8

opaque will not allow light to pass through it 124, 134

organ structure in a plant or animal made of several different *tissues* 9, 15

oscilloscope a device which shows a picture of a wave which represents a sound wave 138, 146

oxidation *oxygen* joining with other *elements* to make *compounds* called oxides; examples are *burning*, rusting and *respiration* 66

oxygen a *gas* that makes up about the *air* 13-16, 19-20, 23-24, 26, . 56-57, 59-61, 63-64, 66, 69-70, 72, ⁊

P

particle a very small piece of matter that everything is made of 51, 54-55, 62, 64-65, 67, 74, 80, 82, 100-103, 107

particle diagram, model a way of picturing matter as made up of moving *particles*; also called the kinetic theory 64-65, 67

penicillin *antibiotic* drug used to kill *bacteria* in the body 30, 36

Periodic Table a table of the *elements* arranged in order so that similar elements are in the same column or group 56, 62

periscope a device, made from two mirrors, which can be used to see over walls or from submarines which are under water 126

pitch how high or low sound is 135-138, 141, 146

plasma the *liquid* part of *blood* 11

pollute, pollution contamination of the *environment* with unwanted *materials* or *energy* 32

population All the plants or animals of one *species* that live in a particular place 32, 39, 43, 46, 48-50

pores the spaces between grains in *porous* rocks 76, 79, 88, 90

porous describes the texture of a rock with *pores* 75-76, 86

predict to say what you think will happen 48

pressure how much pushing *force* there is on an area 72, 87, 91, 93, 95-96

primary colours red, green and blue are the primary colours of light 132

prism   a device, made of glass or plastic, which *refracts* light twice, it can be used to produce a *spectrum*   129-131, 134

producer   a name given to green plants because they produce food   39, 46-47, 50

product   a new substance made in a *chemical reaction*   14, 61, 65-66, 78

proteins   *nutrients* needed for *growth* and repair; made up of *amino acids*   1-3, 5-7, 9, 11-12, 28

pumice   an *igneous rock*; a *lava* with lots of *gas* bubbles   87, 93, 96

pure   contains one *material* only   52, 63, 67, 73-74

pyramid of numbers   pyramid-shaped diagram that shows how the numbers of living things change along a *food chain*   39, 49-50

Q

quadrat   an object, often a square frame, used for sampling living things   39, 43, 46, 50, 149

R

radiation (of heat)   a method of heat transfer, where the *heat energy* is given out as infra-red waves   104, 106, 109

ray diagram   a diagram where light is represented by straight lines   121

ray of light   a beam of light that travels in a straight line   121

react, reaction   what happens when chemicals join or separate   14, 63, 65-66, 70, 74, 77-78, 89

reactant   a substance that you start off with in a *chemical reaction*   14, 61, 65

recycle   use *materials* over and over again   75, 95

red blood cell   a *blood cell* that carries *oxygen*   3-4, 35

reflect, reflected   bounce back, for example light will reflect off a mirror   104-105, 123-126

reflected ray   a *ray of light* which has *reflected* off a mirror   125

refract, refraction, refracted   bending of a *ray of light* when it travels from a *material* of one density to one of a different density   127-129

relay   a device which uses an *electromagnet* to switch on one circuit when a second circuit is complete   118

repel   push apart   110, 112

reproduction   when living things produce young of the same kind as themselves   39

respiration   the breakdown of food to release *energy* in living *cells*   11, 13-15, 19-20, 22, 24, 26, 29, 66

S

saliva   *digestive* juice made in the salivary glands   10, 34

salt   (i) the everyday name for common salt or *sodium chloride*

(ii) one of the *products* of a *reaction* between an *acid* and an *alkali*   59-60, 65, 73

sample   take a small part to get an idea of the whole   43, 147-149

sandstone   a *sedimentary* rock made from sand   75-77, 83, 85-86, 90, 92, 96

scurvy   a disorder caused by lack of *vitamin* C in the diet   4

sediment   rock fragments that settle on the bed of a river, lake or sea   75, 81-84, 86-87, 95-96

sedimentary rocks   rocks formed when *sediments* are compacted and cemented; *sandstone* and *limestone* are examples   75, 77, 83-84, 86-89, 95-96

semi-circular canals   part of the inner ear, used for balance   142

shadow   the far side of an *object* from a light source, so no light falls on the shadow   121

small intestine   the narrow part of the intestine between the *stomach* and the *large intestine*; where *digestion* finishes and *absorption* happens.   7-8, 10-11

Smith, William   1769-1839   85

Snow, John   1813-1858   32-33

sodium chloride   common *salt*; a compound of sodium and chlorine   55, 59-60, 65, 68

sodium hydroxide   a compound that *dissolves* in water to make an *alkali*   24

solid   a substance that stays a definite shape   54, 56, 68 , 71-73, 77-78, 80, 83-84, 86, 91, 93

solidify   change from a *liquid* into a *solid* that happens as a result of cooling   71, 86, 92-95

soluble   able to *dissolve*   7, 9, 78

solution   a *mixture* formed when a *solute dissolves* in a solvent   10, 92

south-seeking pole   the end of a *magnet* which is *attracted* to the Earth's South Pole when allowed to spin freely   112

Spallanzani, Lazzaro   1729-1799   9

species   we say that plants or animals which can interbreed belong to the same species   50

spectrum   white light split into its seven constituent colours, a raindow is an example of a spectrum   129-134

starch   a *carbohydrate* with large, *insoluble molecules*   3, 7, 9-12

stomach   an *organ* in the *digestive system*   8

symbol   a shorthand way of writing the names of *elements*   51, 56-57, 60-62

symbol equation   a *chemical equation* written using *symbols*   61

T

temperature   a measure of the *heat energy* contained in hot objects   91-92, 96-100, 102-104, 106-109

texture   describes rocks as *porous* or *non-porous*   75-76, 86, 88, 90

thermal decomposition   when a *compound* is broken down using heat   66

thermals   hot *air* rising, used by glider pilots to gain height   103

thermometer   a device used to measure *temperature*   97-98, 102, 109

Thermos flask   used to keep hot *liquids* hot or cold liquids cold   106

tinnitus   a hearing defect where the sufferer hears a constant ringing noise   145

tissue   a group of *cells* with the same shape and job   15

tissue fluid   *liquid* between all the *cells* of your body through which *dissolved* substances *diffuse*   15

translucent   allows light to pass through but breaks it up so that there is no clear *image*   123, 134

transmit, transmitted   allows something to pass through   123, 134

transparent   allows light to pass through and gives a clear *image*   123, 134

V

vaccine   substance used to produce *immunity* to a *disease*   27, 37

vacuum   an empty space with nothing inside, not even *air*   106, 139

vascular plants   a group of plants with a specialised transport or vascular system   40-41, 50

vein   *blood* vessels that carry blood towards the *heart*   13, 16-18, 26

vertebrate   an animal with a skeleton made of bone inside its body   39-40, 50

vibrate, vibration   a constant backwards and forwards motion   100, 135-140, 142-146

virus   a *micro-organism* that can only live and reproduce inside living *cells*; a cause of some *infectious diseases*   27-28, 31, 34-36, 38

vitamins   *nutrients* that we need in small amounts to stay healthy   1, 4-5, 7, 11-12

volcanic ash   *igneous rock* formed during explosive volcanic *eruptions*   87, 93, 94, 96

volcano   mountain or hill formed from *lava* or ash during volcanic *eruptions*   87, 93, 96

W

weathering   breakdown of rock caused by rainwater and *temperature* changes   75-84, 86-87, 95-96

white blood cells   *cells* in the *blood* that help to destroy *micro-organisms*; some make *antibodies*   35, 37

word equation   this shows the *reactants* and the *products* of a *chemical reaction* in words   14, 19, 61, 65-66, 89

Y

yeast   a *micro-organism* that we use to make alcohol and bread; a *fungus*   28-30

Acknowledgements

We are grateful to the following for permission to reproduce photographs:

Anthony Blake Photo Library 6t (Phototeque Culinaire) 6c (Eaglemoss Consumer Publications) 6b (Martin Brig); **ArenaPAL.com** 145 (Clive Barda); **British Geological Survey** 75b; **Graham Burns** 49; **Trevor Clifford** 137tr, 137m, 137b; **Corbis** 19tl (JFPI Studios Inc.), 19m (Roger Ressmeyer), 52t (Thom Lang), 60 (Kevin Schafer), 75b (Michael St. Maur Sheil), 81b (Michael Busselle), 84t (Chinch Gryniewicz/Ecoscene), 91tl (Charles O'Rear), 94tr (Yann Arthus-Betrand); **Ecoscene** 42tr (Chinch Gryniewicz), 42br (Chinch Gryniewicz), 44tl (Chinch Gryniewicz), 45l (Sally Morgan), 45bl (Kevin King), 45br (Sally Morgan); **Mary Evans** 9, 18tr; **Geoscience Features Picture Library** 75l, 75r, 76m, 79tl, 79tr, 80t, 84b, 85m, 88l, 88r, 89bl, 90tr, 90ml, 90mr, 90bl, 90br, 92t, 92m, 92b, 93t, 93b, 94bl, 94br; **Robert Harding** 68 (F.Friberg); **Istituto e Museo di Storia della Scienza** (Franca Principe) 130; **Andrew Lambert** 55bl, 77, 78, 89tr; **Nigel Luckhurst** 76t, 89bl, 108; **Jean Martin** 43, 44c, 44r, 149; **Mediscan Medical Images** 10bl; **Vanessa Miles** 22l, 22r, 45mr, 51, 55t, 59, 81mt, 111m, 115, 141b; **Natural History Photographic Agency** 141t (Manfred Danegger), 142t (Alan Williams); **Nature Picture Library** 39 (Bernard Castelein), 142bl (Artur Tabor), 142br (Duncan McEwan); **John Noble** 81t; **Professional Sport** 13r (Tommy Hindley), 23 (Tommy Hindley), **Redferns** 137tl (Geoff Dann), 144m (Mick Hutson); **Science Photo Library** 2r (Peter Menzel), 3tr (Biophotos Associates), 3br (Biophotos Associates), 10r (CNRI), 13l (Cristina Pedrazzini), 14 (Jeremy Walker), 17 (Sheila Terry), 18br (Alfred Pasieka), 19br (James King-Holmes), 20 (Alfred Pasieka), 30 (Maisoneuve Publiphoto Diffusion), 31 (Jane Shemilt), 35 (Noble Proctor), 36 (James King-Holmes), 41l (Dr. Jeremy Burgess), 41r (Th Foto-Werbung), 42c (Lepus), 45t (John Heseltine), 47 (Lepus), 52m (Pascal Goetgheluck), 52b (Claude Nuridsany and Marie Perennou), 55br (Charles D.Winters), 65t (Martyn F. Chillmaid), 65m (Charles D.Winters), 70tl (Tony McConnell), 70tr (Maximilian Stock Ltd), 70br (Simon Fraser), 71 (Charles D.Winters), 72 (David Taylor), 75m (Sinclair Stammers), 76b (Martin Bond), 80b (Sinclair Stammers), 84mt (Matthew Oldfield), 84mb (James King-Holmes), 85t (Nasa), 90tl (George Bernard), 91tr (G. Brad Lewis), 91br (Martin Bond), 93mt (Oscar Burriel), 93mb (Soames Summerhays), 98t (Astrid and Hanns-Frieder Michler), 98tr (Chris Priest and Mark Clarke), 103 (Marty F. Chillmaid), 130 (David Parker), 140 (Keith Kent); **John Walmsley** 136, 144b; **Wellcome Trust Medical Photolibrary** 2l (Fiona Pragoff), 4; **Chris Westwood** 79br, 81mb.

Picture research: Vanessa Miles

The publisher has made every effort to trace copyright holders, but if they have inadvertently overlooked any they will be pleased to make the necessary arrangements at the earliest opportunity.